T0269519

The Birnbaum–Saunders Distribution

The Birnbaum–Saunders Distribution

Víctor Leiva
Adolfo Ibáñez University
Viña del Mar, Chile

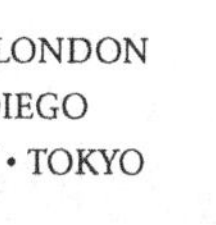

AMSTERDAM • BOSTON • HEIDELBERG • LONDON
NEW YORK • OXFORD • PARIS • SAN DIEGO
SAN FRANCISCO • SINGAPORE • SYDNEY • TOKYO
Academic Press is an imprint of Elsevier

Academic Press is an imprint of Elsevier
125 London Wall, London, EC2Y 5AS, UK
525 B Street, Suite 1800, San Diego, CA 92101-4495, USA
225 Wyman Street, Waltham, MA 02451, USA
The Boulevard, Langford Lane, Kidlington, Oxford OX5 1GB, UK

ISBN: 978-0-12-803769-0

British Library Cataloguing in Publication Data
A catalogue record for this book is available from the British Library

Library of Congress Cataloging-in-Publication Data
A catalog record for this book is available from the Library of Congress

For information on all Academic Press publications
visit our website at http://store.elsevier.com/

I dedicate this book to my PhD students Carolina Marchant, Cristian Villegas, Fabiana Garcia, Fernando Rojas, Helton Saulo, Jeremias Leão, Juan Vega, Manoel Santos, Marcelo Rodríguez, and Michelli Barros, and to my MSc students Camilo Lillo and Edgar Espinosa.

CONTENTS

LIST OF FIGURES

LIST OF TABLES

PREFACE

The objective of this book is to provide an overview of the Birnbaum–Saunders distribution and some of its probabilistic and statistical features, as well as its extension to regression analysis, diagnostics, and goodness of fit. A differentiating aspect of this book is the presentation of codes in the R programming language for analyzing data sets under the Birnbaum–Saunders distribution and related regression models. The R language is a noncommercial and open-source software package for statistical computing and graphics that can be obtained at no cost from http://www.R-project.org.

The genesis of the Birnbaum–Saunders distribution was motivated by problems of vibration in commercial aircrafts causing fatigue in materials. It is also known as the fatigue-life distribution and has been widely applied to fatigue and reliability studies. However, its field of application has been extended beyond the original context of material fatigue and reliability analysis. The Birnbaum–Saunders distribution is now a natural model in many situations where the accumulation of a certain factor forces a quantifiable characteristic to exceed a critical threshold. A few examples of cases in which this distribution can be used are:

(i) Migration of metallic flaws in nano-circuits due to heat in a computer chip.
(ii) Accumulation of deleterious substances in the lungs from air pollution.
(iii) Ingestion by humans of toxic chemicals from industrial waste.
(iv) Occurrence of chronic cardiac diseases and different types of cancer due to cumulative damage caused by several risk factors, which produce degradation and conduct to a fatigue process.
(v) Air quality due to an accumulation of pollutants in the atmosphere over time.
(vi) Generation of action potentials in neural activity.
(vii) Late human mortality due to how the risk of death occurs at a late stage of human life.
(viii) Occurrence of natural disasters such as earthquakes and tsunamis.

All these problems seem to be more amenable to statistical analyses based on the fatigue-life or Birnbaum–Saunders distribution than based on a classic distribution.

The text is organized in six chapters. In Chapter 1, we detail the history of life distributions, how the fatigue processes are developed and the genesis and mathematical derivation of the Birnbaum–Saunders distribution. In Chapter 2, we characterize the more relevant properties and continue to introduce moments and simulation aspects of the Birnbaum–Saunders distribution. In Chapter 3, we discuss procedures to infer and estimate parameters of the Birnbaum–Saunders distribution based on maximum likelihood, moment and graphical methods, considering censored and uncensored data. In Chapter 4, we study regression models with errors following the Birnbaum–Saunders distribution and diagnostic tools for these models. In Chapter 5, we present goodness-of-fit methods and criteria of model selection for the Birnbaum–Saunders distribution. In Chapter 6, we describe several real-world data sets modeled with the Birnbaum–Saunders distribution and analyze them with an implementation in `R` code.

CHAPTER 1

Genesis of the Birnbaum–Saunders Distribution

Abstract

In this chapter, a background and the history of life distributions, as well as some of its indicators, such as the failure rate and reliability function, are provided. In addition, we detail how the fatigue processes are developed and the genesis and mathematical derivation of the Birnbaum–Saunders distribution. Furthermore, a connection between this distribution and the law of proportionate effects is discussed. Moreover, several real-world applications of the Birnbaum–Saunders distribution are mentioned.

Keywords: cumulative damage models, failure or hazard rate, fatigue, law of proportionate effects, life distributions, lifetime data, reliability or survival function.

1.1 INTRODUCTION

In general, the literature related to statistical methods and models for lifetime data recognizes the concept "lifetime" as a positive continuous random variable representing the time until the occurrence of some event of interest. However, in some occasions, the aging of items (e.g., components, systems, subsystems, specimens, structures, organs, units) is not measured in chronological terms. In such occasions, the lifetime is measured by means of other random variables. For example, the amount of kilometers traveled, strength of material specimens until its rupture, level of degradation, flexibility of an adhesive, and number of cycles until a material specimen fails caused by fatigue (fatigue-life). In addition, the terminology "lifetime random variable," which we denote hereafter by T, is widely used for referring to any positive continuous random variable (e.g., amount of rain water or contaminant concentration). Any probabilistic model associated with a lifetime random variable is often called a life distribution. For more details about life distributions, the interested reader is referred to Marshall and Olkin (2007).

The Birnbaum–Saunders Distribution. http://dx.doi.org/10.1016/B978-0-12-803769-0.00001-7

The reliability theory refers to probabilistic and statistical problems, which are related to life distributions of items subject to failure. The probabilistic methods of this theory describe the performance and degradation of items by means of lifetime random variables. Life distributions are used to determinate the reliability (or survival) function, failure (or hazard) rate, and average lifetime of these items. The statistical aspects of this theory solve estimation and inference aspects of the life distribution parameters.

Technically, the reliability of an item is defined as the probability that a lifetime random variable $T > 0$ surpasses a fixed time t. Thus, the reliability function of T over time t is given by

$$R_T(t) = \mathrm{P}(T > t) = 1 - \mathrm{P}(T \leq t) = 1 - F_T(t) = 1 - \int_0^t f_T(u)\,\mathrm{d}u, \quad t > 0, \tag{1.1}$$

which, as noted, depends on the continuous life distribution when a parametric analysis is carried out. This life distribution for T can be characterized by its probability density function $f_T(\cdot)$ or its cumulative distribution function $F_T(\cdot)$, as viewed in Equation (1.1), or even by its failure or hazard rate given by

$$h_T(t) = \frac{f_T(t)}{R_T(t)}, \quad t > 0,$$

or its cumulative failure rate defined as

$$H_T(t) = \int_0^t h_T(u)\,\mathrm{d}u.$$

1.2 HISTORY

The most used probabilistic model is the normal (or Gaussian) distribution. However, often life distributions are two-parameter probabilistic models having asymmetry to the right (positive skewness), unimodality, and positive support ($T > 0$). Thus, the normal distribution is unsuitable for modeling lifetime data. In lifetime parametric analyses, the key distribution is the exponential model, also called the negative exponential distribution. Nevertheless, other probabilistic models have also been used as alternative distributions to the normal model upon asymmetry to the right and positive support. Next, the evolution of life distributions is presented.

Davis (1952) showed several examples of failure data, along with some goodness-of-fit tests for these data. This work provided a guideline toward the use of the exponential model as a life distribution. Later, the works by Epstein and Sobel (1953, 1954a,b, 1955) strengthened and helped to popularize the exponential life distribution. Due to this, the work by Davis (1952) is widely mentioned to justify the hypothesis of an exponential failure distribution, even today. With the publication of the works mentioned, the exponential distribution was transformed into a key reference to model lifetime of items. The exponential probabilistic model, as many other models, is based on a population of finite size. Thus, if this size approaches to infinite, the characteristic properties of the exponential distribution are obtained, emphasizing the property of memory absence. One of the fundamental reasons of the popularity of the exponential distribution, and of its wide applicability in reliability theory, is its constant failure rate (because of its memory absence). Also, its simple data processing, which, in the past, was important due to the computational limitations. The memory absence property of the exponential distribution allows us to simplify problems of inference associated with this model. However, this property is also a disadvantage, since its application to real-world situations is restricted, as is the case of components that age over time.

With the fame of the exponential distribution, the use of the gamma distribution to model the sum of lifetimes with exponential distribution arises. An advantage of the gamma life distribution is that it arises naturally from the convolution of exponentials distributions, but its disadvantage is that the gamma model is not algebraically treatable.

A number of studies began to find lifetime data that are better modeled through a life distribution different to the exponential model. For example, Lieblein and Zelen (1956) and Kao (1959) considered the Weibull distribution, because this model fits failure data well. This distribution was developed by Weibull (1951) to describe the strength to the rupture of materials. The interest for the Weibull life distribution was increasing mainly due to the work by Zelen and Dannemiller (1961), which indicated that many procedures of life tests based on the exponential distribution are unsuitable. The Weibull distribution is very useful, because it is flexible to describe different types of hazard situations and, in addition, it is mathematically tractable.

Motivated by the fatigue of materials present in commercial aircrafts caused by their constant vibration, and the problems provoked by it, Birnbaum and Saunders (1968) derived an ingenious probabilistic model that describes lifetimes associated with material specimens exposed to fatigue, product of cyclical stress and tension. The following year, Birnbaum and Saunders (1969b); Birnbaum and Saunders (1969a) formalized the fatigue-life distribution that later took their names, characterized it as a life distribution and proposed an estimation method for the parameters of this two-parameter distribution. Birnbaum and Saunders (1958) were already interested in finding a model that described the failure time associated with material specimens subject to fatigue. Simultaneously, Birnbaum and Saunders also worked with a group composed by Barlow, Esary, Marshall, and Proschan in wear-out processes caused by cumulative; see Leiva and Saunders (2015). The main results of these investigations were established in the following three works:

(i) Birnbaum et al. (1966) developed a stochastic model of wear-out.
(ii) Birnbaum and Saunders (1969a) derived the fatigue-life distribution.
(iii) Esary et al. (1973) proposed the shock model.

The Birnbaum–Saunders distribution is based on cumulative damage that produces fatigue in the materials. This argument is Miner's rule; see Miner (1945). Birnbaum and Saunders (1968) obtained a probabilistic interpretation of this rule. Thus, the fatigue-life distribution of Birnbaum and Saunders (1969a) was derived from a model showing that the total time elapsed until the cumulative damage, produced by the development and growth of a dominant crack, surpasses a threshold value and causes the material specimen to fail. Desmond (1985) strengthened the justification for the use of this distribution by relaxing some assumptions early made by Birnbaum and Saunders (1969a). In addition, a more general derivation of the Birnbaum–Saunders distribution was provided by Desmond (1985) based on a biological model introduced by Cramér (1946, p. 219). Although the fatigue-life distribution is known as the Birnbaum–Saunders model, a previous work attributed to Freudenthal and Shinozuka (1961) introduced a similar model with a different parameterization. Furthermore, the work by Fréchet (1927) presented some similar ideas related to the Birnbaum–Saunders distribution, but also with a different parameterization to that considered by Birnbaum and Saunders (1969a). This fact was mentioned by Birnbaum and Saunders (1969a) and appears described in Desmond

(1985), Johnson et al. (1995, pp. 651–663), Marshall and Olkin (2007), and Saunders (2007).

After the work of Birnbaum and Saunders (1969a), other probability models were proposed as life distributions in the seventies. Nelson and Hahn (1972) used the log-normal distribution to model lifetime data. This model is commonly employed to describe repair lifetimes in reliability of systems. The repair rate is defined in an analogous way to the failure rate. Similarly, Chhikara and Folks (1977) postulated the inverse Gaussian model as a life distribution, whereas Kalbfleish and Prentice (1980) did the same with the log-logistic model. They were not the only probabilistic models that were used as life distributions. For example, the log-gamma distribution of Bartlett and Kendall (1946), the extreme value distribution of Gumbel (1958), and the truncated normal distribution of Barlow and Proschan (1965) can also be mentioned. However, only some of these models have a genesis from reliability theory or survival analysis.

1.3 FATIGUE, GENESIS, AND DERIVATION

According to a standard terminology (see ASTM International, 2013), fatigue is a process of progressive, localized, and structural change occurring in a material subject to conditions that produce fluctuating levels of stress and strain. This occurs at some point or points of the material and may culminate in cracks or complete fracture after reaching critical levels of stress. Thus, fatigue is a cumulative damage process presented in materials subject to periodic and fluctuating stress levels during each duty cycle. The cumulative damage quantity is the sum of many small damages leading to fatigue of the material, culminating in its failure. Fatigue has been identified as an important cause of failure in engineering structures. When the effect of vibrations on material specimens and structures is studied, the first point considered is the mechanism that could provoke fatigue of these materials.

To understand fatigue processes and later the genesis of the Birnbaum–Saunders distribution, we proceed to recall concepts related to crack, cycle, fatigue, and load. Consider a material specimen exposed to a sequence of loads during a period of time that we call a cycle. Each cycle produces deflection of this specimen, thereby imposing a stress on it. In each cycle, the same sequence of loads is applied to the specimen as in the previous cycle. Suppose that the load is visualized as a continuous unimodal function defined on the unit interval. The value of this function at any time provides the amount of stress imposed to the specimen. Fatigue is the mentioned deflection culminating in structural damage (crack), which is accumulated inside a material specimen over time. The cycles of stress can go from a constant pattern of sinusoidal amplitude to more complex patterns with variations in the amplitude. Failures caused by fatigue take place when the amount of cumulative damage exceeds a threshold level of strength of the material specimen. Generally, the failure is inducted by a fracture that is provoked by a crack in this specimen.

A theory for material fatigue is based on the assumption that the propagation of a dominant crack is the main cause of the failure due to fatigue. In general, a fatigue process can be divided in three stages defined by:

(A1) The beginning of an imperceptible fissure.
(A2) The growth and propagation of the fissure, which provokes a crack in the material specimen due to cyclical stress and tension.
(A3) The rupture or failure of the material specimen due to fatigue.

It is well known that the dominant crack begins between the 5th and 10th percentiles of the service life of the material specimen. The stage (A3) of the fatigue process occupies a negligible lifetime. Therefore, the second stage (A2) of the process contains most of the time of the fatigue life. For this reason, statistical models for fatigue processes are primarily concerned with describing the random variation of lifetimes associated with this second stage through two-parameter life distributions. These parameters allow those material specimens subject to fatigue to be characterized and, at the same time, to predict their behavior under different cyclical stress and tension patterns. In summary, the failure occurs when the total extension of the crack exceeds a critical threshold for the first time. The partial extension of a crack produced by fatigue in each cycle is modeled by a random variable depending on the type of material, magnitude of stress and number of previous cycles, among other factors. For more details about the fatigue process, the reader is referred to Valluri (1963), Birnbaum and Saunders (1969a), Mann et al. (1974, pp. 150–152), Murthy (1974), Saunders (1976), and Volodin (2002).

In practice, the fatigue life for a specimen or structure is the lifetime equivalent to the number of cycles until the failure due to fatigue. For example, fatigue has been recognized as one of the main causes of failure in metal and concrete structures, which is commonly induced by a crack in the material specimen. For modern materials, like carbon fibers and composites, for instance, fatigue is assumed as the development of intangibles produced by concentrations of stress inside the material specimen. Freudental and Gumbel (1954) developed a model that describes the number of cycles until the failure caused by fatigue using the stress level as model factor and so establishing the minimum fatigue life.

As a fatigue-life distribution, the Birnbaum–Saunders model considers a material specimen that is exposed to a sequence of m cyclical loads, denoted by $\{l_i, i = 1, 2, \ldots, m, m \in \mathbb{N}\}$. The loading scheme can be depicted as follows:

$$\underbrace{l_1, l_2, \ldots, l_m}_{\text{1st cycle}} \quad \underbrace{l_{m+1}, l_{m+2}, \ldots, l_{2m}}_{\text{2nd cycle}} \quad \ldots \quad \underbrace{l_{jm+1}, l_{jm+2}, \ldots, l_{jm+m}}_{(j+1)\text{th cycle}},$$

where $l_{jm+i} = l_{km+i}$, for $j \neq k$. Birnbaum and Saunders (1969a) considered that the load is continuous. Thus, at the imposition of each load, l_i, the crack is extended by a random amount. Crack extensions by cycle cannot be observed in practice and only the time instant when the failure occurs is known.

Having explained the physical framework for the genesis of the Birnbaum–Saunders distributions, it is now necessary to make a few statistical assumptions. Birnbaum and Saunders (1969a) used the knowledge of certain types of material failure due to fatigue to derive their model. The fatigue process that they used was based on the following:

(B1) A material is subject to cyclical loads or repetitive shocks, which produce a crack or wear-out in the material.
(B2) The failure in the material occurs when the crack size of the material exceeds a level of resistance (threshold), denoted by ω.
(B3) The sequence of loads imposed to the material is the same from a cycle to another one.
(B4) During a load cycle, the crack extension caused by the loading l_i, X_i, is a random variable whose distribution is governed by all the loads lj, for $j < i$, with $i = 1, \ldots, m$, and by the crack extension that precedes it.
(B5) The total size of the crack caused by the kth cycle, Y_k, is a random variable following a distribution of mean μ and variance σ^2, both of them finite.
(B6) The size of cracks (Y_k) in different cycles are independent.

Note that the crack extension in the $(k+1)$th cycle of load is

$$Y_{k+1} = X_{km+1} + X_{jm+2} + \cdots + X_{km+m}, \quad k = 0, 1, 2, \ldots, m \in \mathbb{N}.$$

As mentioned by Mann et al. (1974, p. 152), assumption (B4) is rather restrictive and may not be valid for some applications. This assumption ensures that, regardless of the dependence among the successive random extensions due to the loads in a particular cycle, the total random crack extensions are independent from cycle to cycle. The plausibility of this assumption in aeronautical fatigue studies is briefly stated by Birnbaum and Saunders (1969a).

The Birnbaum–Saunders model looks for the distribution of the smallest number of cycles, say N^*, such that the sum

$$S_n = \sum_{k=1}^{n} Y_k \tag{1.2}$$

of n positive random variables exceeds the given threshold ω, that is,

$$N^* = \inf\{n \in \mathbb{N} : S_n > \omega\},$$

where S_n is given in Equation (1.2). In the simplest case, the Birnbaum–Saunders distribution is derived by supposing that Y_k are independent identically distributed random variables. Thus, the classic version of the central limit theorem can be applied. In addition, N^* is regarded as a continuous random variable T. Specifically, based on the classic central limit theorem, Equation (1.2) and the assumptions (B1)–(B6), as $n \to \infty$, it is possible to establish that S_n follows approximately a normal distribution ($\dot{\sim}$) with mean $n\mu$ and variance $n\sigma^2$, that is,

$$S_n \dot{\sim} \mathrm{N}(n\mu, n\sigma^2). \tag{1.3}$$

Note that result given in (1.3) corresponds strictly speaking to a convergence in distribution, but we use such a notation for simplicity. Let N be the number of required cycles until the failure of a material. Given that $Y_k > 0$ for all $k \geq 1$, the damage is irreversible and so by complementarity, we have $\{N > n\} \equiv \{S_n \leq \omega\}$ and $\{N \leq n\} \equiv \{S_n > \omega\}$. Then, from assumption (B5) and Equation (1.2), we have that $\mathrm{E}(S_n) = n\mu$ and $\mathrm{Var}(S_n) = n\sigma^2$. Therefore, from Equation (1.3), we get

$$Z_n = \frac{S_n - n\mu}{\sigma\sqrt{n}} \dot{\sim} \mathrm{N}(0, 1). \tag{1.4}$$

Thus, from Equation (1.4) we obtain that

$$\mathrm{P}(N \leq n) \;=\; \mathrm{P}(S_n > \omega)$$

$$= \mathrm{P}\left(Z_n > \frac{\sqrt{\mu\,\omega}}{\sigma}\left(\frac{1}{\sqrt{n}}\sqrt{\frac{\omega}{\mu}} - \sqrt{n}\sqrt{\frac{\mu}{\omega}}\right)\right)$$

$$\approx \Phi\left(\frac{\sqrt{\mu\,\omega}}{\sigma}\left(\sqrt{n}\sqrt{\frac{\mu}{\omega}} - \frac{1}{\sqrt{n}}\sqrt{\frac{\omega}{\mu}}\right)\right), \tag{1.5}$$

where $\Phi(\cdot)$ is the standard normal cumulative distribution function. Birnbaum and Saunders (1969a) used Equation (1.5) to define a continuous life distribution replacing the discrete random variable N by a continuous random variable T, and the discrete argument n by the continuous argument t. Specifically, the number of cycles until the failure N is converted to the total time until the occurrence of a failure T, whereas the nth cycle to the time t. Thus, from Equation (1.5), defining

$$\alpha = \frac{\sigma}{\sqrt{\omega\mu}}, \quad \beta = \frac{\omega}{\mu}, \quad \xi(x) = \sqrt{x} - \frac{1}{\sqrt{x}} = 2\sinh(\log(\sqrt{x})), \quad x > 0,$$

we obtain the cumulative distribution function of the Birnbaum–Saunders distribution with shape parameter $\alpha > 0$ and scale parameter $\beta > 0$ as

$$F_T(t) = \Phi\left(\frac{1}{\alpha}\xi\left(\frac{t}{\beta}\right)\right), \quad t > 0, \ \alpha > 0, \ \beta > 0. \tag{1.6}$$

Note that here and hereafter we use the notation "log" to stand for the natural logarithm. In Saunders (1976), α is also called the "flexure parameter", which is associated directly with the material fatigue. If a random variable T has a cumulative distribution function as in Equation (1.6), then the notation $T \sim \mathrm{BS}(\alpha, \beta)$ is used. This means we are admitting that, if $T \sim \mathrm{BS}(\alpha, \beta)$, it can be written as

$$T = \beta\left(\frac{\alpha}{2}Z + \sqrt{\left(\frac{\alpha}{2}Z\right)^2 + 1}\right)^2, \tag{1.7}$$

where Z is a random variable following the normal distribution with mean $\mu = 0$ and variance $\sigma^2 = 1$, that is, the standard normal distribution, which is denoted by $Z \sim \mathrm{N}(0, 1)$. Hence,

$$Z = \frac{1}{\alpha}\left(\sqrt{\frac{T}{\beta}} - \sqrt{\frac{\beta}{T}}\right) \sim \mathrm{N}(0, 1). \tag{1.8}$$

Therefore, we have two approaches that allow the Birnbaum–Saunders distribution to be obtained. The first approach is originated from physics of materials, which enables the Birnbaum–Saunders model to be interpreted as a life distribution under the assumption that S_n and Y_k represent the total extension of a crack and its partial extension caused by the kth loading cycle, respectively. In this approach, the distribution describes the time elapsed until the crack extension exceeds a threshold conducting to a failure, allowing the Birnbaum–Saunders model to be considered as a life distribution. The second approach is based on Equations (1.7) and (1.8), which is admitted as the definition of the Birnbaum–Saunders distribution. In such an approach, we can assume that any random variable following the Birnbaum–Saunders distribution is a transformation of other random variable with standard normal distribution. From the second approach, some generalizations of the Birnbaum–Saunders distribution can be obtained by switching the random variable Z given in Equation (1.8) with random variables following other distributions. Based on this second approach, we can obtain a huge family of Birnbaum–Saunders distributions having several statistical properties; see Díaz-García and Leiva (2005) and Balakrishnan et al. (2009b). Interestingly, the work introduced by Patriota (2012), which is based on an extended version of the classic central limit theorem, allows us to generalize the Birnbaum–Saunders distribution also using physical arguments. Specifically, the author showed that, under specific persistent dependence structures of the crack extensions, generalizations of the Birnbaum–Saunders distribution can be justified. A number of authors have extended and generalized the Birnbaum–Saunders distribution. The first extension of the Birnbaum–Saunders distribution is attributed to Volodin and Dzhungurova (2000). Then, Díaz-García and Leiva (2005) introduced the generalized Birnbaum–Saunders distribution; see also Leiva et al. (2008c) and Sanhueza et al. (2008b). Owen (2006) proposed a three-parameter extension of the Birnbaum–Saunders distribution. Vilca and Leiva (2006) derived a Birnbaum–Saunders distribution based on skew-normal models. Balakrishnan et al. (2009b) estimated the parameters of the Birnbaum–Saunders distribution with the expectation and maximization algorithm and extended this distribution based on scale-mixture of normal distributions. Gómez et al. (2009) extended the Birnbaum–Saunders distribution from the slash-elliptic model. Guiraud et al. (2009) deducted a noncentral version of the Birnbaum–Saunders distribution. Leiva et al. (2009) provided a length-biased version of the Birnbaum–Saunders distribution. Ahmed et al. (2010) analyzed a truncated version of the Birnbaum–Saunders distribution. Kotz et al. (2010) performed mixture models related to the Birnbaum–Saunders distribution. Vilca et al. (2010) and Castillo et al. (2011) developed the epsilon-skew Birnbaum–Saunders distribution. Balakrishnan et al. (2011) considered Birnbaum–Saunders mixture distributions. Cordeiro and Lemonte (2011) defined the beta-Birnbaum–Saunders distribution. Leiva et al. (2011a) modeled wind energy flux using a shifted Birnbaum–Saunders distribution.

Athayde et al. (2012) viewed the Birnbaum–Saunders distributions as part of the Johnson system, allowing location-scale Birnbaum–Saunders distributions to be obtained. Ferreira et al. (2012) proposed an extreme value version of the Birnbaum–Saunders distribution; see also Gomes et al. (2012). Santos-Neto et al. (2012, 2014) reparameterized the Birnbaum–Saunders distribution obtaining interesting properties. Saulo et al. (2012) presented the Kumaraswamy Birnbaum–Saunders distribution. Fierro et al. (2013) generated the Birnbaum–Saunders distribution from a nonhomogeneous Poisson process. Lemonte (2013) studied the Marshall–Olkin–Birnbaum–Saunders distribution. Bourguignon et al. (2014) derived the power-series Birnbaum–Saunders class of distributions. Cordeiro and Lemonte (2014) proposed the exponentiated generalized Birnbaum–Saunders distribution, Leiva et al. (2015c) introduced a zero-adjusted Birnbaum–Saunders distribution. Bourguignon et al. (2015) derived the transmuted Birnbaum–Saunders distribution.

1.4 APPLICATIONS

Justifying the Birnbaum–Saunders distribution to model data based on an empirical fitting can be a reasonable argument; see details in Chapter 5. However, this argument may be strengthened if we justify theoretically why the Birnbaum–Saunders distribution might be suitable for such a modeling. This theoretical justification can be more attractive to practitioners carrying out applications of the Birnbaum–Saunders distribution in different fields.

Desmond (1985) provided a more general derivation for the Birnbaum–Saunders distribution based on Cramér's (1946) biological model (or law of proportionate effects); see Cramér (1946, p. 219). Desmond (1985) pointed out that the application of Cramér's argument in the fatigue context leads to a Birnbaum–Saunders type distribution, rather than a log-normal distribution, as suggested by Mann et al. (1974, p. 133). Desmond (1985) further provided failure models in random environments described by a stationary continuous-time Gaussian process, for which the Birnbaum–Saunders distribution is more suitable. These models include failures due to the response process being above a fixed level for a long period of time. The damage above a fixed level is related to the so-called exceedance measures of Cramér and Leadbetter (1967). Fatigue failure may also be caused by the stress history, that is proportional to the response of the lightly damped single-degree-of-freedom oscillator excited by a stationary

Gaussian white noise. All of these models lead to Birnbaum–Saunders type distributions.

Let $Y_1 < Y_2 < \cdots < Y_n$ be an ordered sequence of random variables denoting the size of a crack after successive loads. Then, we can write

$$Y_{k+1} = Y_k + U_{k+1}\, g(Y_k), \quad k = 0, 1, \ldots, \tag{1.9}$$

where U_{k+1} is a random variable corresponding to the magnitude of the $(k+1)$th impulse, $g(\cdot) > 0$ a real function and Y_{k+1} the accumulated total amount after application of this impulse.

The relationship defined in Equation (1.9) was originally proposed in a biological context, whereas Desmond (1985) put it in a context of fatigue life. Now, considering Equation (1.9), applying the classic central limit theorem and assuming the increment $\Delta Y_k = Y_{k+1} - Y_k$ in the $(k+1)$th impulse is just small enough to change the summation to integration, we obtain that

$$\sum_{j=1}^{n} U_j = \sum_{j=1}^{n} \frac{\Delta Y_j}{g(Y_j)} \approx \int_{Y_0}^{Y_n} \frac{\mathrm{d}y}{g(y)} \tag{1.10}$$

follows approximately a normal distribution, where Y_0 is the initial crack size. To obtain the fatigue-life or Birnbaum–Saunders distribution, suppose that the mean of U_j is η, its variance ς^2 and the discrete value n is replaced by the continuous value t in Equation (1.10). Then,

$$I(Y_t) = \int_{Y_0}^{Y_t} \frac{1}{g(y)}\,\mathrm{d}y \sim \mathrm{N}(t\eta, t\varsigma^2), \tag{1.11}$$

where Y_t is the crack size at time t. Assume now that $\omega > Y_0$ is the critical crack length at which failure occurs. Then, $T = \inf\{t : Y_t > \omega\}$ is the time to fatigue failure. Thus, from Equation (1.11) and using the equivalent events $\{T \leq t\}$ and $\{Y_t > \omega\}$, it follows that the cumulative distribution function of T at t is given by

$$F_T(t) = \Phi\left(\frac{t\eta - I(\omega)}{\sqrt{t}\varsigma}\right). \tag{1.12}$$

From Equation (1.12), note that the model given in Equation (1.9) leads to life distributions in the Birnbaum–Saunders family independently of the form of $g(Y)$. The choice of the function $g(Y)$ determines the dependence of the rate of crack extension on the previous crack size. Empirical evidence suggests that a power function for $g(Y)$ is reasonable, that is, $g(Y) = Y^{\delta}$, where δ is a material indicator, for $\delta \geq 0$, but $\delta \neq 1$; see Desmond (1985).

Thus, from Equation (1.11), we have

$$I(Y_t) = \int_{Y_0}^{Y_t} \frac{dy}{g(y)} = \frac{Y_0^{1-\delta} - Y_t^{1-\delta}}{\delta - 1} \sim \mathrm{N}(t\eta, t\varsigma^2). \tag{1.13}$$

From Equation (1.13) and by symmetry of the normal distribution, we have

$$Y_t^{1-\delta} \sim \mathrm{N}(Y_0^{1-\delta} + (1-\delta)t\eta, (1-\delta)^2 t\varsigma^2). \tag{1.14}$$

Then, according to Equations (1.12) and (1.14), the cumulative distribution function of T at t is

$$F_T(t) = \begin{cases} \Phi\left(\frac{\omega^{1-\delta} - Y_0^{1-\delta} + (\delta-1)t\eta}{(\delta-1)\sqrt{t}\varsigma}\right), & \text{if } \delta > 1; \\ \Phi\left(\frac{Y_0^{1-\delta} - \omega^{1-\delta} + (1-\delta)t\eta}{(1-\delta)\sqrt{t}\varsigma}\right), & \text{if } \delta < 1. \end{cases} \tag{1.15}$$

From Equation (1.15), note that the Birnbaum–Saunders distribution: (i) results for $\delta = 0$, (ii) can be considered as a transformation of the standard normal distribution, and (iii) has cumulative distribution function such as in (1.6) and given by

$$F_T(t) = \Phi\left(((t/\beta)^{1/2} - (\beta/t)^{1/2})/\alpha\right), \quad \text{where} \quad \alpha = \varsigma/\sqrt{\eta\omega}, \beta = \omega/\eta.$$

From the above derivation, the Birnbaum–Saunders distribution can be widely applied for describing fatigue life and lifetimes in general. However, from this same derivation, its field of application can be extended beyond the original context of material fatigue. The Birnbaum–Saunders distribution may be a natural model in many situations where the accumulation of a certain random variable forces to exceed a critical threshold. Therefore, one can note the versatility of this distribution.

The Birnbaum–Saunders distribution has many applications in a wide variety of contexts; see, for example, Johnson et al. (1995, pp. 651–663). Specifically, some applications of the Birnbaum–Saunders distribution are in failure models in random environments described by stationary Gaussian processes. As mentioned, these models include failures due to response process being above a pre-established level of this response during a long period of time. Also, phenomena to be described by wear-out and cumulative damage processes can be efficiently modeled by the Birnbaum–Saunders distribution. Obviously, based on its genesis, fatigue phenomena, and more extensively lifetimes from wear, abrasion, galling, or wilting, among others, can be modeled ideally by this distribution. Some unpublished applications of the Birnbaum–Saunders distribution to be mentioned are the following:

(i) Migration of metallic flaws in nano-circuits due to heat in a computer chip.
(ii) Accumulation of deleterious substances in the lungs from air pollution.
(iii) Ingestion by humans of toxic chemicals from industrial waste.
(iv) Diminution of biomass in fishing through time in a certain zone.
(v) Occurrence of natural disasters such as earthquakes and tsunamis.

Applications of the Birnbaum–Saunders distribution that have been published outside the fatigue of materials and reliability are the following:

(vi) Analysis of rainfall characteristics of Hiroshima city based on the distribution of periods of continuous rainfall is best fitted by a Birnbaum–Saunders distribution; see Mills (1997), Seto et al. (1993, 1995) and Johnson et al. (1995, p. 655).
(vii) Analysis of airplane components exposed to randomly fluctuating stresses in flight resulting from wind gusts, or to sudden random variations in peak stresses during the take off-landing; see Mills (1997, p. 7) .
(viii) Occurrence of chronic cardiac diseases and different types of cancer due to cumulative damage caused by several risk factors, which produce degradation and conduct to a fatigue process; see Leiva et al. (2007) and Barros et al. (2008).
(ix) Development of intangibles in the execution process of a software that produces cumulative damage deteriorating its performance and ultimately causes its failure; see Balakrishnan et al. (2007).
(x) Disruption of the regeneration process results in death of trees of small diameter at breast height; see Podlaski (2008) and Leiva et al. (2012).
(xi) Risk of contamination in lakes, rivers, and reservoirs from human and agricultural activities, due to vegetal nutrients accumulated over time; see Leiva et al. (2009) and Vilca et al. (2010).
(xii) Wind energy flux and climatology; see Leiva et al. (2011a, 2015a).
(xiii) Air quality due to an accumulation of pollutants in the atmosphere over time; see Leiva et al. (2008a, 2010b, 2015b), Ferreira et al. (2012), Marchant et al. (2013b), and Saulo et al. (2013, 2015).
(xiv) Generation of action potentials in neural activity; see Leiva et al. (2015e).

Another application to be postulated is presented in human aging, where age at death can be adequately described by the Birnbaum–Saunders distribution. In general, its applications are not only limited to the mentioned areas, because recently its use in business, cryptography, economics, finance, industry, insurance, inventory, nutrition, psychology, quality control, and toxicology have also been considered, among others; see Jin and Kawczak (2003), Fox et al. (2008), Lio and Park (2008), Balakrishnan et al. (2009a), Ahmed et al. (2010), Bhatti (2010), Leiva et al. (2010a, 2011b, 2014a,c,d), Lio et al. (2010), Azevedo et al. (2012), Paula et al. (2012), Marchant et al. (2013a), Santos-Neto et al. (2014), Rojas et al. (2015), Sanchez et al. (2015), and Wanke and Leiva (2015). Also, due to its similarity and proximity with the inverse Gaussian distribution, potential applications of the Birnbaum–Saunders distribution can be conducted in actuarial science, agricultural, basic sciences, demography, economy, employment service, engineering, finance, hydrology, labor disputes, linguistics, and toxicology. For more details, see Seshadri (1999b), pp. 167–347), Chhikara and Folks (1989), p. 5), Leiva et al. (2008a,d) and Sanhueza et al. (2008a, 2009, 2011). Applications of cumulative degradation are well-known in engineering, which can also be carried out with the Birnbaum–Saunders distribution; see Meeker and Escobar (1998) and Ho (2012).

Now, as mentioned, the most popular probability model used in statistics is the normal distribution, but not all random phenomena in nature can be described by either the normal or symmetrical distributions. There are many fields in which asymmetrical probability models are used including the above mentioned phenomena. However, limitations of such asymmetric models are found in the availability of suitable statistical methods that are computationally implemented. Practitioners of probability models may refuse the use of distributions that are different to the normal and exponential ones, because they must become familiar with a theory that is new to them. This also makes difficult the applicability of new models such as the Birnbaum–Saunders distribution, in spite of their potential benefits. Nevertheless, efforts for developing basic research in statistics can be more attractive to practitioners if the new developed models and methods are computationally implemented and available, for example, in the R software; see www.r-project.org and R-Team (2015). The Birnbaum–Saunders distribution is implemented in the R software by the bs and gbs packages; see Leiva et al. (2006) and Barros et al. (2009), respectively. The bs and gbs packages can be downloaded from https://cran.r-project.org/src/contrib/Archive/bs and https://cran.r-project.org/src/contrib/Archive/gbs, respectively, or from http://www.victorleiva.cl.

CHAPTER 2

Characterizations of the Birnbaum–Saunders Distribution

Abstract

In this chapter, the more relevant features and properties, moments, and simulation aspects of the Birnbaum–Saunders distribution are presented, several of them by using its relation with the standard normal distribution. Specifically, we detail probabilistic features, such as the probability density, cumulative distribution and quantile functions, as well as lifetime indicators, such as the reliability or survival function and the failure or hazard rate. From the probability density and quantile functions, we are able to determine the mode and the median of the distribution. Note that the failure rate of the Birnbaum–Saunders distribution has a unimodal shape. Then, its change point and limiting behavior are discussed. In addition. we describe several ways to obtain the moments of the Birnbaum–Saunders distribution, including moments of negative, fractional and positive orders. Furthermore, its characteristic function is proposed and used for determining the corresponding moments. Moreover, by using the relation between the Birnbaum–Saunders and log-Birnbaum–Saunders distributions and the Bessel function, we obtain moments of all orders, including negative moments. Particularly, the mean, variance and coefficients of variation, skewness (asymmetry) and kurtosis are provided. This chapter is finished detailing several generators of random numbers for the Birnbaum–Saunders distribution using the methods of multiple roots and probability integral transform, as well as the relations between the Birnbaum–Saunders distribution and the inverse Gaussian, log-Birnbaum–Saunders and standard normal distributions.

Keywords: Bessel function, binomial theorem, characteristic function, cumulative distribution function, failure or hazard rate, gamma function, inverse Gaussian distribution, moments, multiple roots and probability integral transform methods, probability density function, quantile function, reliability or survival function, standard normal distribution.

The Birnbaum–Saunders Distribution. http://dx.doi.org/10.1016/B978-0-12-803769-0.00002-9

2.1 INTRODUCTION

We recall that, if a random variable T follows this distribution with shape parameter $\alpha > 0$ and scale parameter $\beta > 0$, then the notation $T \sim \text{BS}(\alpha, \beta)$ is used. From Equation (1.8) in Chapter 1 given by

$$Z = \frac{1}{\alpha}\left(\sqrt{\frac{T}{\beta}} - \sqrt{\frac{\beta}{T}}\right),$$

the random variables $T \sim \text{BS}(\alpha, \beta)$ and $Z \sim \text{N}(0,1)$ are related by a monotone transformation. Thus, as mentioned in Chapter 1, we conclude that any random variable T with Birnbaum–Saunders distribution can be obtained as a transformation of another random variable Z with standard normal distribution. Then, we obtain that

$$T = \beta\left(\frac{\alpha Z}{2} + \sqrt{\left(\frac{\alpha Z}{2}\right)^2 + 1}\right)^2. \tag{2.1}$$

2.2 PROBABILITY FUNCTIONS AND PROPERTIES

Let $T \sim \text{BS}(\alpha, \beta)$. Then, the probability density function of T is given by

$$\begin{aligned} f_T(t; \alpha, \beta) &= \frac{1}{\sqrt{2\pi}} \exp\left(-\frac{1}{2\alpha^2}\left(\frac{t}{\beta} + \frac{\beta}{t} - 2\right)\right) \frac{1}{2\alpha\beta}\left(\left(\frac{t}{\beta}\right)^{-1/2} + \left(\frac{t}{\beta}\right)^{-3/2}\right) \\ &= \frac{1}{\sqrt{8\pi}} \exp\left(\frac{1}{\alpha^2}\right) \exp\left(-\frac{1}{2\alpha^2}\left(\frac{t}{\beta} + \frac{\beta}{t}\right)\right) \frac{t^{-3/2}}{\alpha\beta^{1/2}}(t + \beta), \end{aligned} \tag{2.2}$$

for $t > 0$, $\alpha > 0$, and $\beta > 0$.

The expression given in Equation (2.2) can be easily obtained using the transformation theorem of random variables considering

$$Z = \frac{1}{\alpha}\left(\sqrt{\frac{T}{\beta}} - \sqrt{\frac{\beta}{T}}\right) \sim \text{N}(0, 1), \tag{2.3}$$

whose probability density function is given by

$$\phi(z) = \frac{1}{\sqrt{2\pi}} \exp\left(-\frac{1}{2}z^2\right), \quad z \in \mathbb{R}. \tag{2.4}$$

Thus, based on Equation (2.3) and using the transformation theorem of random variables, the probability density function of $T \sim \text{BS}(\alpha, \beta)$ is given by

$$f_T(t; \alpha, \beta) = \phi\left(\frac{1}{\alpha}\xi\left(\frac{t}{\beta}\right)\right)\left|\frac{\mathrm{d}}{\mathrm{d}t}\left(\frac{1}{\alpha}\xi\left(\frac{t}{\beta}\right)\right)\right|$$
$$= \frac{1}{\sqrt{2\pi}}\exp\left(-\frac{1}{2}\left(\frac{1}{\alpha}\xi\left(\frac{t}{\beta}\right)\right)^2\right)\frac{1}{\alpha}\xi'\left(\frac{t}{\beta}\right), \quad t > 0, \quad (2.5)$$

where

$$\xi(u) = u^{1/2} - u^{-1/2} = \sinh(\log(u)), \quad u > 0, \quad (2.6)$$

and the derivative of the function $\xi(\cdot)$ given in Equation (2.6) is

$$\xi'(u) = \frac{\mathrm{d}}{\mathrm{d}u}\xi(u) = \frac{1}{2}\left(u^{-1/2} + u^{-3/2}\right), \quad u > 0.$$

The expression obtained in Equation (2.5) proves the result defined in Equation (2.2). The mode of $T \sim \text{BS}(\alpha, \beta)$, denoted by t_m, is given by the solution of $(\beta - t_m)(t_m + \beta)^2 = \alpha^2\beta t_m(t_m + 3\beta)$.

Figure 2.1 displays graphical plots of the probability density function of the Birnbaum–Saunders distribution for different values of its shape parameter α, considering its scale parameter $\beta = 1$ (without loss of generality), without loss of generality. Based on this figure, note that the Birnbaum–Saunders distribution is continuous, unimodal and positively

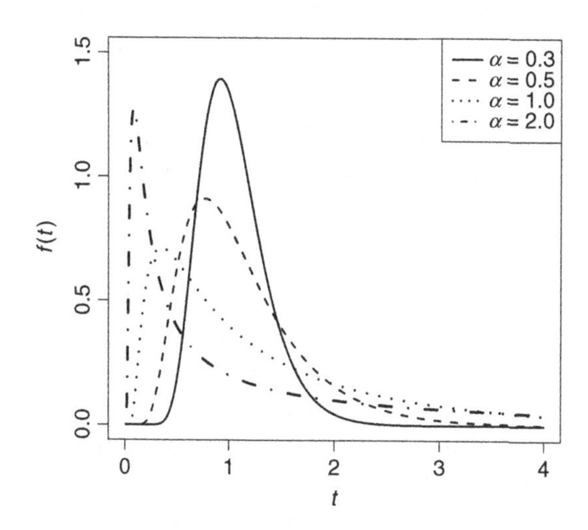

Figure 2.1 Plots of the Birnbaum–Saunders probability density function for the indicated value of α with $\beta = 1.0$.

skewed (asymmetry to right). It is also possible to note that, as α tends to zero, the Birnbaum–Saunders distribution tends to be symmetrical around β (the median of the distribution) and its variability decreases. Furthermore, notice that, as the shape parameter increases, the Birnbaum–Saunders distribution has heavier tails. This means that α modifies the skewness and kurtosis of the distribution.

From the monotone transformation given in Equation (2.1), we can obtain the cumulative distribution function of $T \sim \mathrm{BS}(\alpha, \beta)$ as

$$F_T(t; \alpha, \beta) = \mathrm{P}(T \leq t) = \mathrm{P}\left(\beta\left(\frac{\alpha Z}{2} + \sqrt{\left(\frac{\alpha Z}{2}\right)^2 + 1}\right)^2 \leq t\right)$$
$$= \mathrm{P}\left(Z \leq \frac{1}{\alpha}\left(\sqrt{\frac{t}{\beta}} - \sqrt{\frac{\beta}{t}}\right)\right).$$

Therefore,

$$F_T(t; \alpha, \beta) = \Phi\left(\frac{1}{\alpha}\xi\left(\frac{t}{\beta}\right)\right), \quad t > 0, \ \alpha > 0, \ \beta > 0, \tag{2.7}$$

where $\xi(\cdot)$ is given in Equation (2.6) and

$$\Phi(z) = \int_{-\infty}^{z} \phi(u)\, \mathrm{d}u, \quad z \in \mathbb{R}, \tag{2.8}$$

is the standard normal cumulative distribution function, with $\phi(\cdot)$ given in Equation (2.4). Note that the function defined in Equation (2.7) can also be obtained by integrating the probability density function defined in Equation (2.5).

Another useful indicator for statistical analyses is the quantile function or $q \times 100$th quantile of the distribution. Based on Equation (2.7), the quantile function of $T \sim \mathrm{BS}(\alpha, \beta)$ is given by

$$t(q; \alpha, \beta) = F_T^{-1}(p; \alpha, \beta) = \beta\left(\frac{\alpha z(q)}{2} + \sqrt{\left(\frac{\alpha z(q)}{2}\right)^2 + 1}\right)^2, \tag{2.9}$$

for $0 < q \leq 1$, where $F_T^{-1}(\cdot)$ is the inverse function of $F_T(\cdot)$ expressed in Equation (2.7) and $z(q) = \Phi^{-1}(q)$ is the $q \times 100$th quantile of the standard normal distribution, with $\Phi^{-1}(\cdot)$ being the inverse function of $\Phi(\cdot)$ given in Equation (2.8). Note that if $q = 0.5$, then $z(0.5) = 0$, which corresponds to the mean, median, and mode of $Z \sim \mathrm{N}(0, 1)$. Thus, from the quantile function defined in Equation (2.9), we have that $t(0.5; \alpha, \beta) = \beta$, which confirms that the scale parameter $\beta > 0$ is also the median of the Birnbaum–Saunders distribution. Equation given in Equation (2.9) can be used for generating random numbers in simulation processes of the Birnbaum–Saunders distribution and also for deriving goodness-of-fit tools associated with it.

Two useful indicators in lifetime analysis are the reliability or survival function and the failure or hazard rate. From Equations (2.5) and (2.7), the reliability function and failure rate of $T \sim \mathrm{BS}(\alpha, \beta)$ are, respectively, given by

$$R_T(t; \alpha, \beta) = 1 - F_T(t; \alpha, \beta) = \Phi\left(-\frac{1}{\alpha}\xi\left(\frac{t}{\beta}\right)\right), \quad t > 0, \tag{2.10}$$

$$h_T(t; \alpha, \beta) = \frac{f_T(t; \alpha, \beta)}{R_T(t; \alpha, \beta)} = \frac{\phi\left(\frac{1}{\alpha}\xi\left(\frac{t}{\beta}\right)\right)\frac{1}{\alpha}\xi'\left(\frac{t}{\beta}\right)}{\Phi\left(-\frac{1}{\alpha}\xi\left(\frac{t}{\beta}\right)\right)}, \quad t > 0, \tag{2.11}$$

for $0 < R_T(t; \alpha, \beta) < 1$. Thus, we have the following three statements for the failure rate of the Birnbaum–Saunders distribution defined in Equation (2.11):

(C1) $h_T(t; \alpha, \beta)$ is unimodal for any α, increasing for $t < t_c$, and decreasing for $t > t_c$, where t_c is the change-point of $h_T(t; \alpha, \beta)$.
(C2) $h_T(t; \alpha, \beta)$ approaches $1/(2\alpha^2\beta)$ as $t \to \infty$.
(C3) $h_T(t; \alpha, \beta)$ tends to be increasing as $\alpha \to 0$.

For more details about (C1)–(C3), see Chang and Tang (1993) and Kundu et al. (2008). Numerical studies indicate that the Birnbaum–Saunders distribution has approximately an increasing failure rate when $\alpha < 0.41$ and $0 < t < 8\beta$, which implies an increasing failure rate in average. Birnbaum and Saunders (1969a) showed by numerical calculations that the average failure rate of T decreases slowly for $t < 1.64$.

Figure 2.2 displays graphical plots of the failure rate of the Birnbaum–Saunders distribution for different values of α and $\beta = 1$ (without loss of generality). Theoretical considerations described in (C1)–(C3) are verified by this graphical shape analysis.

The Birnbaum–Saunders distribution has proportionality and reciprocity properties, that is, the distribution belongs to the scale and closed under reciprocation families; see Saunders (1974) and Marshall and Olkin (2007).

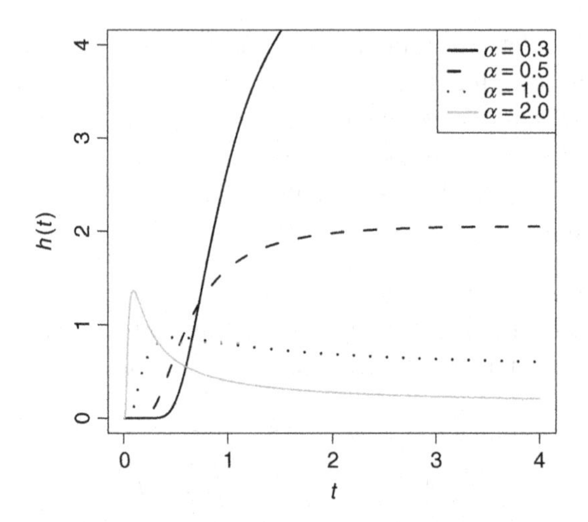

Figure 2.2 Plots of the Birnbaum–Saunders failure rate for the indicated value of α *with* $\beta = 1.0$.

From the Birnbaum–Saunders probability density function given in Equation (2.2), clearly if $T \sim \text{BS}(\alpha, \beta)$, for all $b > 0$, the random variable $Y = bT$ follows a Birnbaum–Saunders distribution with parameters α and $b\beta$. In addition, the random variable $Y = 1/T$ has the same distribution of T with the parameter β replaced by $1/\beta$. Specifically, let $T \sim \text{BS}(\alpha, \beta)$. Then,

(D1) $bT \sim \text{BS}(\alpha, b\beta)$ for $b > 0$.
(D2) $\frac{1}{T} \sim \text{BS}\left(\alpha, \frac{1}{\beta}\right)$.

In addition, from the representation given in Equation (2.3), note that any random variable following a Birnbaum–Saunders distribution is related to the chi-squared distribution with one degree of freedom and mean equal to one and variance equal to two. Thus, from Equation (2.3), a further property of the Birnbaum–Saunders distribution is:

(D3) $V = Z^2 = \frac{1}{\alpha^2}\left(\frac{T}{\beta} + \frac{\beta}{T} - 2\right) \sim \chi^2(1)$, with $\text{E}(V) = 1$ and $\text{Var}(V) = 2$.

Furthermore, other important representation of a random variable following the Birnbaum–Saunders distribution is related to the its logarithmic transformation. Hence, if $T \sim \text{BS}(\alpha, \beta)$, then:

(D4) $Y = \log(T) \sim \text{log-BS}(\alpha, \log(\beta))$, where $\text{log-BS}(\alpha, \log(\beta))$ stands for a distribution known as logarithmic Birnbaum–Saunders, which is presented in Chapter 4. We call it as the log-Birnbaum–Saunders distribution.

Properties (D1)–(D4) of the Birnbaum–Saunders distribution are useful for diverse statistical purposes, such as generation of moments and of random numbers, estimation of parameters, and modeling based on regression; see more details in Saunders (1974), Rieck and Nedelman (1991), and Rieck (1999) and in the next sections of this chapter and in Chapter 3.

2.3 CHARACTERISTIC FUNCTION AND MOMENTS

Let T be a continuous random variable with probability density function $f_T(\cdot)$. Then, the characteristic function of T is defined by

$$\eta_T(s) = \mathrm{E}(\exp(\mathrm{i}sT)) = \int_0^\infty \exp(\mathrm{i}st) f_T(t)\,\mathrm{d}t,$$

where $\exp(\mathrm{i}sT) = \cos(sT) + \mathrm{i}\sin(sT)$, with $\mathrm{i} = \sqrt{-1}$ being the imaginary unit. The characteristic function is also known as the Fourier transform of $f_T(\cdot)$. Thus, once the characteristic function is determined, we can establish the probability density function of T and vice versa. It is also possible to determine the moments of rth order of the random variable T through the relation

$$\mathrm{E}(T^r) = \frac{\eta_T^{(r)}(0)}{\mathrm{i}^r}, \tag{2.12}$$

where $\eta_T^{(r)}(0)$ denotes the rth derivative of the characteristic function $\eta_T(\cdot)$ evaluated at $r = 0$.

Let $T \sim \mathrm{BS}(\alpha, \beta)$. Then, the characteristic function of T is obtained by using

$$\eta_T(s) = \int_0^\infty \exp(\mathrm{i}st) \frac{1}{\sqrt{2\pi}} \exp\left(-\frac{1}{2\alpha^2}\left(\frac{t}{\beta} + \frac{\beta}{t} - 2\right)\right) \times \frac{1}{2\alpha\beta}\left(\left(\frac{t}{\beta}\right)^{-1/2} + \left(\frac{t}{\beta}\right)^{-3/2}\right) \mathrm{d}t.$$

Thus, from above and for $\beta = 1$, we have

$$\eta_T(s) = \frac{1}{2\sqrt{2\pi}} \int_0^\infty \frac{1}{\alpha}\left(t^{-1/2} + t^{-3/2}\right) \exp\left(-\frac{1}{2\alpha^2 t} - \left(\frac{1}{2\alpha^2} - \mathrm{i}s\right) t + \frac{1}{\alpha^2}\right) \mathrm{d}t$$

$$= \frac{\exp(\frac{1}{\alpha^2})}{\sqrt{2\pi}\alpha}\left(\frac{1}{2}\int_0^\infty t^{1/2-1}\exp\left(-\left(\frac{1-2\mathrm{i}s\alpha^2}{2\alpha^2}\right)t-\frac{1}{2\alpha^2 t}\right)\mathrm{d}t\right.$$
$$\left.+\frac{1}{2}\int_0^\infty t^{-1/2-1}\exp\left(-\left(\frac{1-2\mathrm{i}s\alpha^2}{2\alpha^2}\right)t-\frac{1}{2\alpha^2 t}\right)\mathrm{d}t\right).$$

Considering in the expression above $x = 1 - 2\mathrm{i}s\alpha^2$ and the variable change $v = (x/(2\alpha^2))t$, we have

$$\eta_T(s) = \frac{\exp(\frac{1}{\alpha^2})}{\sqrt{2\pi}\alpha}\left(x^{1/4}\frac{1}{2}\left(\frac{\sqrt{x}}{\alpha^2}\right)^{-1/2}\int_0^\infty v^{1/2-1}\exp\left(-v-\frac{(\sqrt{x}/\alpha^2)^2}{4v}\right)\mathrm{d}v\right.$$
$$\left.+x^{-1/4}\frac{1}{2}\left(\frac{\sqrt{x}}{\alpha^2}\right)^{1/2}\int_0^\infty v^{-1/2-1}\exp\left(-v-\frac{(\sqrt{x}/\alpha^2)^2}{4v}\right)\mathrm{d}v\right).$$
$$= \frac{\exp(\frac{1}{\alpha^2})}{\sqrt{2\pi}\alpha}\left(x^{1/4}K_{-1/2}\left(\frac{\sqrt{x}}{\alpha^2}\right)+x^{-1/4}K_{1/2}\left(\frac{\sqrt{x}}{\alpha^2}\right)\right),$$

where $K_\lambda(\cdot)$ is the modified Bessel function of the third kind detailed in Equations (2.42), (2.44)-(2.49), and (2.56). Specifically, we use the result $K_{1/2}(u) = K_{-1/2}(u) = \exp(-u)\sqrt{\pi/(2u)}$ to find the characteristic function of $T \sim \mathrm{BS}(\alpha,\beta)$ given by

$$\eta_T(s) = \frac{1}{2}\left(\left(1+\frac{1}{\sqrt{1-2\,\mathrm{i}\,s\,\alpha^2\,\beta}}\right)\exp\left(\frac{1-\sqrt{1-2\,\mathrm{i}\,s\,\alpha^2\,\beta}}{\alpha^2}\right)\right). \quad (2.13)$$

The rth moment of the continuous random variable T is given in general by

$$\mathrm{E}(T^r) = \int_0^\infty t^r f_T(t)\,\mathrm{d}t, \quad (2.14)$$

where $f_T(\cdot)$ is the probability density function of T. Thus, the existence of moments depends on the convergence of the integral defined in Equation (2.14). In the case of the Birnbaum–Saunders distribution, all positive and negative moment exist; see Rieck (1999). The positive integer moments can be obtained by taking the derivative of the characteristic function given in Equation (2.13) according to Equation (2.12). Alternatively, in order to find the moments of the Birnbaum–Saunders distribution, we can use the relation established in Equation (2.1). Hence, the expected value of the random variable $(T/\beta)^r$ can be expressed as

$$\mathrm{E}\left(\left(\frac{T}{\beta}\right)^r\right) = \mathrm{E}\left(\left(\frac{\alpha Z}{2}+\sqrt{\left(\frac{\alpha Z}{2}\right)^2+1}\right)^{2r}\right). \quad (2.15)$$

From Equation (2.15) and the binomial theorem defined as

$$(a+b)^m = \sum_{k=0}^{m} \binom{m}{k} a^{m-k} b^k,$$

we obtain

$$\mathrm{E}\left(\left(\frac{T}{\beta}\right)^r\right) = \sum_{k=0}^{2r} \binom{2r}{k} \mathrm{E}\left(\left(\left(\frac{\alpha Z}{2}\right)^2 + 1\right)^{k/2} \left(\frac{\alpha Z}{2}\right)^{2r-k}\right).$$

Now, note that

$$\mathrm{E}\left(\left(\left(\frac{\alpha Z}{2}\right)^2 + 1\right)^t \left(\frac{\alpha Z}{2}\right)^s\right) = 0,$$

if s is odd. Therefore,

$$\mathrm{E}\left(\left(\frac{T}{\beta}\right)^r\right) = \sum_{k=0}^{r} \binom{2r}{2k} \mathrm{E}\left(\left(\left(\frac{\alpha Z}{2}\right)^2 + 1\right)^k \left(\frac{\alpha Z}{2}\right)^{2(r-k)}\right).$$

By expanding the binomial expression

$$\left(\left(\frac{\alpha Z}{2}\right)^2 + 1\right)^j = \sum_{k=0}^{j} \binom{j}{k} \left(\frac{\alpha Z}{2}\right)^{2k}, \tag{2.16}$$

we obtain that

$$\begin{aligned} \mathrm{E}\left(\frac{T}{\beta}\right)^r &= \sum_{j=0}^{r} \binom{2r}{2j} \mathrm{E}\left(\sum_{k=0}^{j} \binom{j}{k} \left(\frac{\alpha Z}{2}\right)^{2k} \left(\frac{\alpha Z}{2}\right)^{2(r-j)}\right) \\ &= \sum_{j=0}^{r} \binom{2r}{2j} \sum_{k=0}^{j} \binom{j}{k} \left(\frac{\alpha}{2}\right)^{2(r-j+k)} \mathrm{E}(Z^{2(r-j+k)}). \end{aligned} \tag{2.17}$$

Now, as $Z \sim \mathrm{N}(0, 1)$, then

$$\mathrm{E}\left(Z^{2(r-j+k)}\right) = \frac{(2(r-j+k))!}{2^{r-j+k}(r-j+k)!};$$

see Roussas (1997, p. 116). Therefore,

$$\mathrm{E}\left(\left(\frac{T}{\beta}\right)^r\right) = \sum_{j=0}^{r} \binom{2r}{2j} \sum_{k=0}^{j} \binom{j}{k} \frac{(2(r-j+k))!}{2^{r-j+k}(r-j+k)!} \left(\frac{\alpha}{2}\right)^{2(r-j+k)}. \tag{2.18}$$

The expression given in Equation (2.18) permits us to obtain the moments of the Birnbaum–Saunders distribution. Through these moments, it is possible to find the mean, variance, and coefficients of skewness and kurtosis of the distribution.

To find the mean of $T \sim \mathrm{BS}(\alpha, \beta)$, we can take $r = 1$ in Equation (2.18), obtaining

$$\begin{aligned}
\mathrm{E}\left(\frac{T}{\beta}\right) &= \sum_{j=0}^{1} \binom{2}{2j} \sum_{k=0}^{j} \binom{j}{k} \frac{(2(1-j+k))!}{2^{1-j+k}(1-j+k)!} \left(\frac{\alpha}{2}\right)^{2(1-j+k)} \\
&= \binom{2}{0} \frac{2!}{2\,1!} \left(\frac{\alpha}{2}\right)^{2} + \binom{2}{2} \sum_{k=0}^{1} \binom{1}{k} \frac{(2k)!}{2^k k!} \left(\frac{\alpha}{2}\right)^{2k} \\
&= 1 + \frac{\alpha^2}{2}.
\end{aligned}$$

Then,

$$\mathrm{E}(T) = \beta \left(1 + \frac{\alpha^2}{2}\right). \tag{2.19}$$

To find the variance of $T \sim \mathrm{BS}(\alpha, \beta)$, since $\mathrm{Var}(T) = \mathrm{E}(T^2) - (\mathrm{E}(T))^2$, we evaluate Equation (2.18) with $r = 2$ obtaining

$$\begin{aligned}
\mathrm{E}\left(\left(\frac{T}{\beta}\right)^2\right) &= \sum_{j=0}^{2} \binom{4}{2j} \sum_{k=0}^{j} \binom{j}{k} \frac{(2(2-j+k))!}{2^{2-j+k}(2-j+k)!} \left(\frac{\alpha}{2}\right)^{2(2-j+k)} \\
&= 3\left(\frac{\alpha}{2}\right)^4 + 6\left(\frac{\alpha}{2}\right)^2 + 18\left(\frac{\alpha}{2}\right)^4 + 1 + 2\left(\frac{\alpha}{2}\right)^2 + 3\left(\frac{\alpha}{2}\right)^4 \\
&= \frac{3}{2}\alpha^4 + 2\alpha^2 + 1.
\end{aligned}$$

Therefore,

$$\mathrm{E}(T^2) = \beta^2 \left(\frac{3}{2}\alpha^4 + 2\alpha^2 + 1\right) \tag{2.20}$$

and then

$$\begin{aligned}
\mathrm{Var}(T) &= \beta^2 \left(\frac{3}{2}\alpha^4 + 2\alpha^2 + 1\right) - \left(\beta\left(1 + \left(\frac{\alpha^2}{2}\right)\right)\right)^2 \\
&= \beta^2\alpha^2 \left(1 + \frac{5}{4}\alpha^2\right).
\end{aligned} \tag{2.21}$$

To find the coefficient of variation of $T \sim \mathrm{BS}(\alpha, \beta)$, given in general by

$$\mathrm{CV}(T) = \gamma(T) = \gamma = \frac{\sigma}{\mu}, \tag{2.22}$$

where $\sigma = \sqrt{\mathrm{Var}(T)}$ and $\mu = \mathrm{E}(T)$, it is enough to replace Equations (2.19) and (2.21) in Equation (2.22), obtaining

$$\gamma(T) = \alpha\left(\frac{\sqrt{5\alpha^2+4}}{\alpha^2+2}\right), \tag{2.23}$$

which does not depend on the scale parameter β. The coefficients of skewness and kurtosis of a random variable T are given in general by

$$\mathrm{CS}(T) = \alpha_3(T) = \sqrt{\beta_1(T)} = \frac{\mu_3}{(\mu_2)^{3/2}}, \tag{2.24}$$

$$\mathrm{CK}(T) = \alpha_4(T) = \beta_2(T) = \frac{\mu_4}{(\mu_2)^2}, \tag{2.25}$$

respectively, where $\mu_k = \mathrm{E}((T - \mathrm{E}(T))^k)$, $k = 2, 3, 4$, and $\sigma = \sqrt{\mu_2}$, according to Johnson et al. (1993, p. 42)'s notation. More generally, $\alpha_k(T) = \mu_k(\mu_2)^{-k/2}$. Thus, to find the coefficients of skewness and kurtosis of $T \sim \mathrm{BS}(\alpha, \beta)$, we need to obtain $\mathrm{E}(T^3)$ and $\mathrm{E}(T^4)$, that is, we must evaluate Equation (2.18) at $r = 3$ and $r = 4$, respectively. Then, if $r = 3$ and developing similarly to $r = 1$ and $r = 2$, we get

$$\begin{aligned}\mathrm{E}\left(\left(\frac{T}{\beta}\right)^3\right) &= \sum_{j=0}^{3}\binom{6}{2j}\sum_{i=0}^{j}\binom{j}{i}\frac{(2(2-j+i))!}{2^{2-j+i}(2-j+i)!}\left(\frac{\alpha}{2}\right)^{2(2-j+i)}\\ &= 1 + \frac{9}{2}\alpha^2 + 9\alpha^4 + \frac{15}{2}\alpha^6,\end{aligned}$$

and if $r = 4$, we have

$$\begin{aligned}\mathrm{E}\left(\left(\frac{T}{\beta}\right)^4\right) &= \sum_{j=0}^{4}\binom{8}{2j}\sum_{i=0}^{j}\binom{j}{i}\frac{(2(2-j+i))!}{2^{2-j+i}(2-j+i)!}\left(\frac{\alpha}{2}\right)^{2(2-j+i)}\\ &= 1 + 8\alpha^2 + 30\alpha^4 + 60\alpha^6 + \frac{105}{2}\alpha^8.\end{aligned}$$

Therefore,

$$\mathrm{E}(T^3) = \frac{\beta^3}{2}\left(2 + 9\alpha^2 + 18\alpha^4 + 15\alpha^6\right), \tag{2.26}$$

$$\mathrm{E}(T^4) = \frac{\beta^4}{2}\left(2 + 16\alpha^2 + 60\alpha^4 + 120\alpha^6 + 105\alpha^8\right). \tag{2.27}$$

Based on Equations (2.19), (2.20), and (2.24)–(2.27), we obtain that the coefficients of skewness and kurtosis of $T \sim \mathrm{BS}(\alpha, \beta)$ are, respectively, given by

$$\mathrm{CS}(T) = \alpha_3(T) = \sqrt{\beta_1(T)} = \frac{16\alpha^2(11\alpha^2+6)^2}{(5\alpha^2+4)^3}, \tag{2.28}$$

$$\mathrm{CK}(T) = \alpha_4(T) = \beta_2(T) = 3 + \frac{6\alpha^2(93\alpha^2 + 40)}{(5\alpha^2 + 4)^2}. \tag{2.29}$$

Negative moments of the Birnbaum–Saunders distribution can be obtained in the following way. If $T \sim \mathrm{BS}(\alpha, \beta)$, then T/β and β/T have the same distribution, that is,

$$\frac{T}{\beta} \sim \mathrm{BS}(\alpha, 1), \quad \frac{\beta}{T} \sim \mathrm{BS}(\alpha, 1). \tag{2.30}$$

Thus, based on result given in Equation (2.30), we obtain

$$\mathrm{E}(T^{-r}) = \frac{\mathrm{E}(T^r)}{\beta^{2r}}. \tag{2.31}$$

Therefore, we can get the mean and the variance of T^{-1} by

$$\mathrm{E}\left(\frac{1}{T}\right) = \frac{1}{\beta}\left(1 + \frac{\alpha^2}{2}\right), \tag{2.32}$$

$$\mathrm{Var}\left(\frac{1}{T}\right) = \frac{\alpha^2}{\beta^2}\left(\frac{5}{4}\alpha^2 + 1\right). \tag{2.33}$$

Now, as the coefficients of variation, skewness and kurtosis do not depend on the scale parameter β, then, for the random variable $1/T$, they coincide with those from the random variable T, that is,

$$\mathrm{CV}\left(\frac{1}{T}\right) = \gamma\left(\frac{1}{T}\right) = \gamma(T) = \alpha\left(\frac{\sqrt{5\alpha^2 + 4}}{\alpha^2 + 2}\right), \tag{2.34}$$

$$\mathrm{CS}\left(\frac{1}{T}\right) = \alpha_3\left(\frac{1}{T}\right) = \sqrt{\beta_1\left(\frac{1}{T}\right)} = \sqrt{\beta_1(T)} = \frac{16\alpha^2(11\alpha^2 + 6)^2}{(5\alpha^2 + 4)^3}, \tag{2.35}$$

$$\mathrm{CK}\left(\frac{1}{T}\right) = \alpha_4\left(\frac{1}{T}\right) = \beta_2\left(\frac{1}{T}\right) = \beta_2(T) = 3 + \frac{6\alpha^2(93\alpha^2 + 40)}{(5\alpha^2 + 4)^2}. \tag{2.36}$$

An interesting result for the negative moments of $T \sim \mathrm{BS}(\alpha, \beta)$ is related to Property (D3). Thus, since $V = Z^2 = (1/\alpha^2)(T/\beta + \beta/T - 2) \sim \chi^2(1)$, then $\mathrm{E}(V) = 1$ and $\mathrm{Var}(V) = 2$, implying

$$\mathrm{E}\left(\frac{T}{\beta} + \frac{\beta}{T} - 2\right) = \alpha^2, \tag{2.37}$$

$$\mathrm{Var}\left(\frac{T}{\beta} + \frac{\beta}{T} - 2\right) = 2\alpha^4. \tag{2.38}$$

Therefore, from the coefficient of variation associated with Equations (2.37) and (2.38), we have

$$\frac{\sqrt{2\,\mathrm{Var}\left(\frac{T}{\beta}+\frac{\beta}{T}-2\right)}}{\mathrm{E}\left(\frac{T}{\beta}+\frac{\beta}{T}-2\right)} = 1. \tag{2.39}$$

Rieck (1999) developed an alternative procedure for generating the moments of a random variable with Birnbaum–Saunders distribution. This procedure is particularly useful for obtaining fractional moments. It is based on the notion that, for a non-negative random variable T, as those with Birnbaum–Saunders distribution for example, the moment generating function can be obtained through the logarithm of T, that is, $Y = \log(T)$. Thus, the rth moment of $T \sim \mathrm{BS}(\alpha, \beta)$ is given by

$$M_Y(r) = \mathrm{E}(\exp(rY)) = \mathrm{E}(T^r). \tag{2.40}$$

Therefore, the random variable $T \sim \mathrm{BS}(\alpha, \beta)$ has moments of all order, that is, negative, fractional, and positive.

As mentioned in Property (D4), the logarithm of $T \sim \mathrm{BS}(\alpha, \beta)$ follows a log-Birnbaum–Saunders distribution; see Rieck and Nedelman (1991). The moment generating function of a random variable $Y = \log(T) \sim$ log-$\mathrm{BS}(\alpha, \log(\beta))$ is expressed as

$$M_Y(r) = \exp(r\ \log(\beta))\left(\frac{K_{2r+1/2}(1/\alpha^2) + K_{2r-1/2}(1/\alpha^2)}{2K_{1/2}(1/\alpha^2)}\right), \tag{2.41}$$

where $K_\lambda(\cdot)$ is the modified Bessel function of the third kind given by

$$K_\lambda(u) = \frac{1}{2}\left(\frac{u}{2}\right)^\lambda \int_0^\infty v^{-\lambda-1} \exp\left(-v - \frac{u^2}{4v}\right) \mathrm{d}v; \tag{2.42}$$

see Gradshteyn and Randzhik (2000, p. 907). Using the relation between the Birnbaum–Saunders and log-Birnbaum–Saunders distributions, the expected value of the random variable T^r, with $T \sim \mathrm{BS}(\alpha, \beta)$, from Equations (2.40) and (2.41), is expressed by

$$\mathrm{E}(T^r) = \beta^r\left(\frac{K_{r+1/2}(1/\alpha^2) + K_{r-1/2}(1/\alpha^2)}{2K_{1/2}(1/\alpha^2)}\right). \tag{2.43}$$

Let us consider the following results for the modified Bessel function of the third kind given in Equation (2.42):

$$K_{1/2}(u) = \sqrt{\frac{\pi}{2u}}\exp(-u), \tag{2.44}$$

$$K_{3/2}(u) = \sqrt{\frac{\pi}{2u}}\exp(-u)(1 + 1/u), \tag{2.45}$$

$$K_{5/2}(u) = \sqrt{\frac{\pi}{2u}}\exp(-u)(1 + 3/u + 3/u^2), \tag{2.46}$$

$$K_{\nu+1/2}(u) = \sqrt{\frac{\pi}{2y}}\exp(-u)\sum_{k=0}^{\nu}\frac{(\nu + k)!}{k!(\nu - k)!(2k)^k}, \tag{2.47}$$

$$K_\nu(u) = K_{-\nu}(u). \tag{2.48}$$

Rieck (1999) also presented the following approximation for large values of u:

$$K_\nu(u) = \sqrt{\frac{\pi}{2u}}\exp(-u)\left(1 + \frac{4\nu^2 - 1}{8u} + \frac{(4\nu^2 - 1)(4\nu^2 - 9)}{2!(8u)^2} + \frac{(4\nu^2 - 1)(4\nu^2 - 9)(4\nu^2 - 25)}{3!(8u)^3} + \cdots\right). \tag{2.49}$$

The approximation given in Equation (2.49) has the property that, if ν is a non-negative real number and u is positive, the remainder after k terms of this approximation does not exceed the $(k + 1)$th term in absolute value and has its same sign for $k > \nu - 1/2$. Note that Gradshteyn and Randzhik (2000) provided expressions for integer order Bessel functions. Alternatively, these functions can be calculated with the `R` software by means of the command `besselK()` or from the function list in Excel.

Based on the results of the Bessel functions provided in Equations (2.44)–(2.48) and (2.43), it can be demonstrated that

$$\mathrm{E}(T) = \beta\left(\frac{K_{3/2}(1/\alpha^2) + K_{1/2}(1/\alpha^2)}{2K_{1/2}(1/\alpha^2)}\right) = \beta\left(1 + \frac{\alpha^2}{2}\right), \tag{2.50}$$

$$\mathrm{E}(T^2) = \beta^2\left(\frac{K_{5/2}(1/\alpha^2) + K_{3/2}(1/\alpha^2)}{2K_{1/2}(1/\alpha^2)}\right) = \beta^2\left(\frac{3}{2}\alpha^4 + 2\alpha^2 + 1\right). \tag{2.51}$$

Therefore,

$$\mathrm{Var}(T) = \beta^2\alpha^2\left(\frac{5}{4}\alpha^2 + 1\right). \tag{2.52}$$

Expressions given in Equations (2.50)–(2.52) coincide with whose given in Equations (2.19)–(2.21), respectively. A more general expression for $\mathrm{E}(T^r)$ based on Equations (2.43) and (2.49) is provided by

$$\mathrm{E}(T^r) = \beta^r \left(\sum_{k=0}^{r} \frac{\alpha^{2k}(r+k)!}{k!(r-k)!2^k} + \sum_{k=0}^{r-1} \frac{\alpha^{2k}(r+k-1)!}{k!(r-k-1)!2^k} \right). \tag{2.53}$$

Based on Equation (2.43), it is also possible to obtain moments of negative order. For example,

$$\mathrm{E}\left(\frac{1}{T}\right) = \frac{1}{\beta}\left(\frac{K_{-1/2}(1/\alpha^2) + K_{-3/2}(1/\alpha^2)}{2K_{1/2}(1/\alpha^2)}\right) = \frac{1}{\beta}\left(1 + \frac{\alpha^2}{2}\right),$$

which coincides with Equation (2.32). However, the most interesting contribution of the procedure for generating moments of the Birnbaum–Saunders distribution proposed by Rieck (1999) is related to the calculation of fractional moments, which are not possible to obtain from Equation (2.18). Thus, moments of fractional order of the Birnbaum–Saunders distribution can be easily obtained from Equation (2.43). For example,

$$\mathrm{E}(T^{1/2}) = \beta^{1/2}\left(\frac{K_1(1/\alpha^2) + K_0(1/\alpha^2)}{2K_{1/2}(1/\alpha^2)}\right), \tag{2.54}$$

$$\mathrm{E}(T^{3/2}) = \beta^{3/2}\left(\frac{K_2(1/\alpha^2) + K_1(1/\alpha^2)}{2K_{1/2}(1/\alpha^2)}\right). \tag{2.55}$$

To find $K_\nu(u)$ when ν is an integer number, we can use the expression

$$K_\nu(u) = \frac{\pi}{2}\left(\frac{I_{-\nu}(u) - I_\nu(u)}{\sin(\nu\pi)}\right), \quad |\arg(u)| < \pi, \tag{2.56}$$

where

$$I_\nu(u) = \sum_{k=0}^{\infty} \frac{1}{k!\,\Gamma(k+\nu+1)}\left(\frac{u}{2}\right)^{\nu+2k}, \quad |u| < \infty, \quad |\arg(u)| < \pi,$$

which is given in Abramowitz and Stegun (1972, p. 375), with $\Gamma(\cdot)$ being the usual gamma function given by

$$\Gamma(u) = \int_0^\infty x^{u-1}\exp(-x)\mathrm{d}x.$$

Alternatively, the approximations

$$K_0(u) \approx \log(u) \quad \text{and} \quad K_\nu(u) \approx \frac{(\nu-1)!}{2}\left(\frac{u}{2}\right)^{-\nu}$$

can be used; see Press et al. (1992, p. 236). In this way,

$$K_0(1/\alpha^2) \approx -2\log(\alpha), \tag{2.57}$$

$$K_1(1/\alpha^2) \approx \frac{1}{4\alpha^2}, \tag{2.58}$$

$$K_2(1/\alpha^2) \approx \frac{1}{8\alpha^4}. \tag{2.59}$$

Then, replacing Equations (2.44), (2.57), and (2.58) in Equations (2.54) and (2.55), we get

$$\mathrm{E}(T^{1/2}) \approx \beta^{1/2}\frac{1}{\sqrt{2\pi}\,\alpha}\exp(\alpha^2)\left(1 - 8\alpha^2\log(\alpha)\right), \tag{2.60}$$

$$\mathrm{E}(T^{3/2}) \approx \beta^{3/2}\exp(-\alpha^2)\frac{1}{2\alpha^5\sqrt{2\pi}}\left(1 + 2\alpha^2\right). \tag{2.61}$$

Proceeding similarly to Equations (2.60) and (2.61), expressions for other fractional orders can be obtained. In Rieck (1999, Table 1), a list of expected values is presented for the square root and power $3/2$ of a random variable with Birnbaum–Saunders distribution for values of α from 0.1 to 1 by 0.1, when $\beta = 1$.

In Rieck and Nedelman (1991), it is pointed out that in practice α usually takes small values, for example, less than 0.5. In Birnbaum and Saunders (1969a), three fatigue data sets were analyzed and the estimates of α were less than 0.5. For small values of the parameter α, as previously indicated, and based on the approximation defined by Rieck (1999) and presented in Equation (2.49), it is possible to provide an approximate expression for the expected value of $T \sim \mathrm{BS}(\alpha, \beta)$ given by

$$\mathrm{E}(T^r) \approx \beta^r\left(1 + \frac{(\alpha r)^2}{2} + \frac{\alpha^4(r^4 - r^2)}{8} + \frac{\alpha^6(r^6 - 5r^4 + 4r^2)}{48}\right). \tag{2.62}$$

The interested reader is referred to Rieck (1999) for details about the precision of the approximation provided in Equation (2.62).

2.4 GENERATION OF RANDOM NUMBERS

The process of mathematical simulation basically consists of constructing a model that recreates the essential aspects of a phenomenon. A simulation is also useful for designing and conducting experiments employing the simulation model, extracting conclusions from its results that support decision-making.

Chang and Tang (1994b) provided the following method for generating random numbers for the Birnbaum–Saunders distribution. Let $T \sim \mathrm{BS}(\alpha, \beta)$. Then, we recall its cumulative distribution function given in Equation (2.7) is

$$F_T(t; \alpha, \beta) = \Phi\left(\frac{1}{\alpha}\left(\sqrt{\frac{t}{\beta}} - \sqrt{\frac{\beta}{t}}\right)\right), \quad t > 0, \ \alpha > 0, \ \beta > 0, \quad (2.63)$$

which implies that, as mentioned in (D3),

$$V = \frac{1}{\alpha^2}\left(\frac{T}{\beta} + \frac{\beta}{T} - 2\right) \quad (2.64)$$

follows a $\chi^2(1)$ distribution. Thus, from Equation (2.64), we have

$$T^2 - \beta(2 + \alpha V)T + \beta^2 = 0. \quad (2.65)$$

Note that Equation (2.65) has two solutions, the roots t_1 and t_2, where $t_1 t_2 = \beta^2$ and $t_2 = \beta^2/t_1$.

Michael et al. (1976) presented a method for generating random variables through transformations with multiple roots. This method can be adapted to the Birnbaum–Saunders distribution because it consists of two continuous random variables, say V and T, related by $V = g(T)$. Michael et al.'s method uses relation of the type $V = g(T)$, which has k roots denoted by $t_1, t_2, \ldots, t_k$, one of which corresponds to a value t_0 with probability

$$p_i(t_0) = \left(1 + \sum_{j=1 \neq i}^{k} \left|\frac{g'(t_i)}{g'(t_j)}\right| \frac{f_T(t_j)}{f_T(t_i)}\right)^{-1}, \quad i = 1, \ldots, k, \quad (2.66)$$

where $g'(\cdot)$ is the derivative of $g(\cdot)$ given above, $|x|$ the absolute value of x, and $f_T(\cdot)$ the probability density function of T. For the Birnbaum–Saunders distribution, we have $k = 2$ and the probability $p_1(t_0) = 1 - p_2(t_0)$ given in Equation (2.66), from which we must choose the smallest root, say t_1, being it equal to 1/2. This is obtained from Equation (2.66), where, considering $t_1 t_2 = \beta^2$ and $t_2 = \beta^2/t_1$, we have

$$\frac{g'(t_1)}{g'(t_2)} = \frac{\frac{1}{\beta} - \frac{\beta}{t_1^2}}{\frac{1}{\beta} - \frac{\beta}{t_2^2}} = \frac{\frac{1}{\beta} - \frac{\beta}{t_1^2}}{\frac{1}{\beta} - \frac{t_1^2}{\beta^3}} = -\left(\frac{\beta}{t_1}\right)^2 \quad (2.67)$$

and

$$\frac{f_T(t_1)}{f_T(t_2)} = \frac{(t_2 + \beta)}{t_2^{3/2}} \frac{t_1^{3/2}}{(t_1 + \beta)} \quad (2.68)$$

$$\times \exp\left(-\frac{1}{2\alpha^2}\left(\left(\sqrt{\frac{t_2}{\beta}}-\sqrt{\frac{\beta}{t_2}}\right)^2+\left(\sqrt{\frac{t_1}{\beta}}-\sqrt{\frac{\beta}{t_1}}\right)^2\right)\right)=\left(\frac{t_1}{\beta}\right)^2.$$

Therefore, from Equations (2.67) and (2.68), and based on Equations (2.66), it is proven that $p_1(t_0) = 1 - p_2(t_0) = 1/2$. Thus, a Birnbaum–Saunders distributed random variable with shape parameter α and scale parameter β can be generated by

$$t_1 1_{[-\infty,0.5]}(u) + t_2 1_{[0.5,\infty]}(u), \tag{2.69}$$

where $1_A(\cdot)$ is the indicator function of the set A. Observe that the expression given in Equation (2.69), according to Michael et al. (1976), assures that the probability of choosing one of the two roots t_1 or t_2, given that $V = v$, is equal to $1/2$. Michael et al.'s procedure for generating random numbers is known as the multiple root method and is described for the Birnbaum–Saunders distribution by Algorithm 1.

Algorithm 1 Generator 1 of random numbers from a Birnbaum–Saunders distribution

1: Generate a random number u from $U \sim \mathrm{U}(0, 1)$.
2: Generate a random number z from $Z \sim \mathrm{N}(0, 1)$.
3: Set values for α and β of $T \sim \mathrm{BS}(\alpha, \beta)$.
4: Compute a random number $t = t_1$ or $t = t_2$ from $T \sim \mathrm{BS}(\alpha, \beta)$ with the following criterion:
4.1: If $u \leq 0.5$, then

$$t_1 = \frac{(\beta(2+\alpha^2 z^2)) - \sqrt{(-\beta(2+\alpha^2 z^2)^2 - 4\beta^2)}}{2};$$

4.2: Else, that is, if $u > 0.5$,

$$t_2 = \frac{(\beta(2+\alpha^2 z^2)) + \sqrt{(-\beta(2+\alpha^2 z^2)^2 - 4\beta^2)}}{2}.$$

5: Repeat steps 1 to 4 until the required amount of random numbers to be completed.

In addition to the multiple root method for generating random numbers from the Birnbaum–Saunders distribution proposed by Chang and Tang (1994b), Rieck (2003) introduced a method for generating random numbers based on the relation between the Birnbaum–Saunders and log-Birnbaum–Saunders distributions; see Rieck and Nedelman (1991) and Property (D4). If $T \sim \mathrm{BS}(\alpha, \beta)$, then $Y = \log(T)$ has a cumulative distribution function given by

$$F_Y(y) = \Phi\left(\frac{2}{\alpha}\sinh\left(\frac{y-\log(\beta)}{2}\right)\right), \quad y \in \mathbb{R}, \ \alpha > 0, \ \log(\beta) \in \mathbb{R}. \tag{2.70}$$

Thus,

$$Z = \left(\frac{2}{\alpha}\sinh\left(\frac{Y-\log(\beta)}{2}\right)\right) \sim \mathrm{N}(0,1). \tag{2.71}$$

The solution of Equation (2.71) for Y is expressed as

$$Y = \log(\beta) + 2\log\left(\frac{\alpha Z}{2} + \left(\frac{\alpha^2 Z^2}{4} + 1\right)^{1/2}\right). \tag{2.72}$$

Then, using the relation between the Birnbaum–Saunders and log-Birnbaum–Saunders distributions, a generator of random numbers for the Birnbaum–Saunders distribution is obtained by taking $\exp(Y)$ in Equation (2.72), which is summarized in Algorithm 2.

Algorithm 2 Generator 2 of random numbers from a Birnbaum–Saunders distribution

1: Generate a random number z from $Z \sim \mathrm{N}(0,1)$.
2: Set values for α and β of $T \sim \mathrm{BS}(\alpha, \beta)$.
3: Compute a random number y from $Y = \log(T) \sim \text{log-BS}(\alpha, \log(\beta))$ by using Equation (2.72) conducting to

$$y = \log(\beta) + 2\log\left(\frac{\alpha z}{2} + \left(\frac{\alpha^2 z^2}{4} + 1\right)^{1/2}\right).$$

4: Calculate a random number $t = \exp(y)$ from $T \sim \mathrm{BS}(\alpha, \beta)$ by using the value y obtained in step 3.
5: Repeat step 1 to 4 until the required amount of random numbers to be completed.

A third method for generating random numbers from a Birnbaum–Saunders model is based on the relation between the Birnbaum–Saunders and inverse Gaussian distributions; see Bhattacharyya and Fries (1982), Desmond (1986), and Balakrishnan et al. (2009a). For more details about the inverse Gaussian distribution, see the books by Chhikara and Folks (1989) and Seshadri (1999a,b). Let $X_1 \sim \mathrm{IG}(\beta, \beta/\alpha), X_2 \sim \mathrm{IG}(1/\beta, 1/(\alpha^2\beta))$, and $Y = 1/X_2$. Then, the random variable

$$T = WX_1 + (1 - W)Y \tag{2.73}$$

follows a Birnbaum–Saunders distribution with parameters α and β, where

$$\mathrm{P}(W = 0) = \mathrm{P}(W = 1) = \frac{1}{2}.$$

Thus, using the relation between the Birnbaum–Saunders and inverse Gaussian models given in Equation (2.73), a generator of random numbers for the Birnbaum–Saunders distribution is summarized in Algorithm 3. In Chhikara and Folks (1989, p. 53), it is possible to find a method for generating random numbers from an inverse Gaussian distribution.

Algorithm 3 Generator 3 of random numbers from a Birnbaum–Saunders distribution

1: Set values for α and β of $T \sim \mathrm{BS}(\alpha, \beta)$.
2: Generate a random number x_1 from $X_1 \sim \mathrm{IG}(\beta, \beta/\alpha)$.
3: Simulate a random number x_2 from $X_2 \sim \mathrm{IG}(1/\beta, 1/(\alpha^2\beta))$.
4: Compute a random number $y = 1/x_2$.
5: Obtain a random number w according to the probability distribution of W given by

$$\mathrm{P}(W = 0) = \mathrm{P}(W = 1) = 1/2.$$

6: Calculate a random number t from $T \sim \mathrm{BS}(\alpha, \beta)$ by using

$$t = wx_1 + (1 - w)y.$$

7: Repeat steps 1 to 6 until the required amount of random numbers to be completed.

The methods based on (i) multiple roots, (ii) the log-Birnbaum–Saunders distribution, and (iii) the inverse Gaussian distribution have some advantages ones on the others. However, the procedure most used for generating random numbers from a probability model consists of inverting the cumulative distribution function, called the inverse transform method (or analogously the probability integral transform). As it is known, if a random variable T has a cumulative distribution function $F_T(\cdot)$, then the transformation of the random variable T given by

$$U = F_T(T) \tag{2.74}$$

follows a uniform distribution in the interval [0, 1], that is, $U = F_T(T) \sim \mathrm{U}(0, 1)$. The transformation $U = F_T(T)$ defined in Equation (2.74) is known as probability integral transform. Hence, the random variable $T = F_T^{-1}(U)$ follows a probability model with cumulative distribution function $F_T(\cdot)$, where $F_T^{-1}(\cdot)$ is the inverse function of $F_T(\cdot)$. Keeping this in mind, it is possible to obtain random numbers from the interest distribution with an adequate generator of random numbers following a uniform distribution in the interval [0, 1].

Let $T \sim \mathrm{BS}(\alpha, \beta)$. Then, its quantile function (or inverse function of the cumulative distribution function) is given in Equation (2.9). Note that the Birnbaum–Saunders quantile function depends on the quantile function of the standard normal distribution, which does not have a closed analytical form. However, it is possible to assume that the Birnbaum–Saunders quantile function has a closed analytical form, which we recall is given by

$$t(q; \alpha, \beta) = \beta \left(\frac{\alpha z(q)}{2} + \sqrt{\frac{\alpha^2 z(q)^2}{4} + 1} \right)^2, \quad 0 < q < 1, \tag{2.75}$$

where $z(q)$ is the $q \times 100$th quantile of the standard normal distribution. Then, using the Birnbaum–Saunders quantile function, that is, the inverse transform method, a generator of random numbers for the Birnbaum–Saunders distribution is summarized in Algorithm 4.

Algorithm 4 Generator 4 of random numbers from a Birnbaum–Saunders distribution

1: Generate a random number z from $Z \sim \mathrm{N}(0, 1)$.
2: Set values for α and β of $T \sim \mathrm{BS}(\alpha, \beta)$.
3: Compute a random number t from $T \sim \mathrm{BS}(\alpha, \beta)$ by using Equation (2.75) conducting to

$$t = \beta \left(\frac{\alpha z}{2} + \sqrt{\frac{\alpha^2 z^2}{4} + 1} \right)^2.$$

4: Repeat steps 1 to 3 until the required amount of random numbers to be completed.

Note from Algorithm 4 that the inverse transform method for generating random numbers from the Birnbaum–Saunders distribution seems to be simpler than the other three generators defined in Algorithms 1, 2, and 3, and then more efficient computationally than them. The interested reader is referred to Leiva et al. (2008d) for a simulation study about the computational efficiency of the generators of random numbers from the Birnbaum–Saunders distribution here presented.

CHAPTER 3

Inference for the Birnbaum–Saunders Distribution

Abstract

In this chapter, a review about estimation methods for the parameters of the Birnbaum–Saunders distribution is provided. We detail the maximum likelihood method for estimating these parameters with complete (uncensored) and censored data. In addition, we discuss a method of modified moments for such an estimation. This chapter is finished describing a graphical approach, which can be used for estimating the corresponding parameters by means of the least square method used in regression. This graphical approach may also be employed as a goodness-of-fit method for the Birnbaum–Saunders distribution. Asymptotic inference is also considered for all of these methods.

Keywords: asymptotic inference, censored data, empirical distribution function, failure rate, least square method, maximum likelihood method, moment estimation method, probability density function, quantile function, regression, reliability function, standard normal distribution.

3.1 INTRODUCTION

Several types of estimation methods for the parameters α and β of the Birnbaum–Saunders distribution have been discussed. Birnbaum and Saunders (1969b) found the maximum likelihood estimates of α and β and proposed the mean–mean estimate as an initial estimate for β. As mentioned by Bhattacharyya and Fries (1982), the lack of an exponential family structure makes it difficult to develop statistical inference procedures for these parameters. Engelhardt et al. (1981), Ahmad (1988), Achcar (1993), and Dupuis and Mills (1998) have discussed other estimators for α and β. However, in all these cases, it is not possible to find explicit expressions for the estimators and numerical procedures must be used.

The Birnbaum–Saunders Distribution. http://dx.doi.org/10.1016/B978-0-12-803769-0.00003-0

A method for estimating α and β based on modified moment estimation, which provides easy analytical expressions to compute the estimates of α and β, was introduced by Ng et al. (2003). From and Li (2006) also presented and summarized several estimation methods for the Birnbaum–Saunders distribution. The interested reader is referred to Mills (1997), Dupuis and Mills (1998), Lemonte et al. (2007, 2008), and Cysneiros et al. (2008), for some results on statistical inference for the Birnbaum–Saunders distribution. Next, we discuss the estimation of the parameter $\boldsymbol{\theta} = (\alpha, \beta)^\top$ of the Birnbaum–Saunders distribution with three methods. The interested reader is referred to the above-mentioned references for other estimation methods.

3.2 MAXIMUM LIKELIHOOD ESTIMATION METHOD

Let $T_1, T_2, \ldots, T_n$ be a random sample of size n from a random variable T with probability density function $f_T(\cdot)$ and $t_1, t_2, \ldots, t_n$ their observations (data). Then, due to the independence among $T_1, T_2, \ldots, T_n$, the likelihood function for a parameter $\boldsymbol{\theta}$ is defined in general as

$$L(\boldsymbol{\theta}) = \prod_{i=1}^{n} L_i(\boldsymbol{\theta}), \tag{3.1}$$

where $L_i(\boldsymbol{\theta}) = L(\boldsymbol{\theta}; t_i) = f_T(t_i; \boldsymbol{\theta})$ is the individual contribution of each observation to the likelihood function. However, as it is well known, because the likelihood function for $\boldsymbol{\theta}$ given in Equation (3.1) and its natural logarithm (log-likelihood hereafter) attain their maximum values at the same points, one often prefers to work with the log-likelihood function due to its mathematical treatability. Hence, the log-likelihood function for $\boldsymbol{\theta}$ is given in general by

$$\ell(\boldsymbol{\theta}) = \sum_{i=1}^{n} \ell_i(\boldsymbol{\theta}), \tag{3.2}$$

where $\ell_i(\boldsymbol{\theta}) = \log(f_T(t_i; \boldsymbol{\theta}))$ is the individual contribution of each observation to the log-likelihood function.

Let $T_1, T_2, \ldots, T_n$ be a random sample of size n from $T \sim \mathrm{BS}(\alpha, \beta)$ and $t_1, t_2, \ldots, t_n$ their observations. Then, the log-likelihood function for $\boldsymbol{\theta} = (\alpha, \beta)^\top$ based on $t_1, t_2, \ldots, t_n$ is given by

$$\ell(\boldsymbol{\theta}) = c_1 + \frac{n}{\alpha^2} - \frac{1}{2\alpha^2}\sum_{i=1}^{n}\left(\frac{t_i}{\beta}+\frac{\beta}{t_i}\right) - n\log(\alpha) - \frac{n}{2}\log(\beta) + \sum_{i=1}^{n}\log(t_i+\beta), \tag{3.3}$$

where c_1 is a constant that does not depend on $\boldsymbol{\theta}$.

To maximize the log-likelihood function $\ell(\boldsymbol{\theta}) = \ell(\alpha,\beta)$ expressed in Equation (3.3), we need its first derivatives with respect to α and β forming the score vector defined by $\dot{\boldsymbol{\ell}} = (\dot{\ell}_\alpha, \dot{\ell}_\beta)^\top$, whose elements are given by

$$\dot{\ell}_\alpha = \frac{\partial \ell(\alpha,\beta)}{\partial\alpha} = -\frac{2n}{\alpha^3} + \frac{1}{\alpha^3}\sum_{i=1}^{n}\left(\frac{t_i}{\beta}+\frac{\beta}{t_i}\right) - \frac{n}{\alpha}, \tag{3.4}$$

$$\dot{\ell}_\beta = \frac{\partial \ell(\alpha,\beta)}{\partial\beta} = \frac{1}{2\alpha^2}\sum_{i=1}^{n}\left(\frac{t_i}{\beta^2}-\frac{1}{t_i}\right) - \frac{n}{2\beta} + \sum_{i=1}^{n}\frac{1}{t_i+\beta}.$$

Thus, to obtain the log-likelihood equations, we make $\dot{\ell}_\alpha = 0$ and $\dot{\ell}_\beta = 0$ and evaluate them at $\alpha = \hat{\alpha}$ and $\beta = \hat{\beta}$, getting

$$\hat{\alpha} = \left(\frac{1}{n}\sum_{i=1}^{n}\left(\frac{t_i}{\hat{\beta}}+\frac{\hat{\beta}}{t_i}\right)\right)^{1/2}, \tag{3.5}$$

$$\hat{\beta} = \sqrt{\frac{\frac{1}{2\hat{\alpha}^2}\sum_{i=1}^{n} t_i}{\frac{1}{2\hat{\alpha}^2}\sum_{i=1}^{n}\frac{1}{t_i} - \sum_{i=1}^{n}\frac{1}{t_i+\hat{\beta}} + \frac{n}{2\hat{\beta}}}}. \tag{3.6}$$

Hence, by using a numerical algorithm, we can solve Equations (3.5) and (3.6). This algorithm allows us to have the joint iterative procedure

$$\hat{\alpha}^{(m+1)} = \left(\frac{1}{n}\sum_{i=1}^{n}\left(\frac{t_i}{\hat{\beta}^{(m)}}+\frac{\hat{\beta}^{(m)}}{t_i}-2\right)\right)^{1/2}, \quad m = 0,1,2,\ldots, \tag{3.7}$$

$$\hat{\beta}^{(m+1)} = \sqrt{\frac{\frac{1}{2(\hat{\alpha}^{(m)})^2}\sum_{i=1}^{n} t_i}{\frac{1}{2(\hat{\alpha}^{(m)})^2}\sum_{i=1}^{n}\frac{1}{t_i} - \sum_{i=1}^{n}\frac{1}{t_i+\hat{\beta}^{(m)}} + \frac{n}{2\hat{\beta}^{(m)}}}},$$

which needs starting values $\hat{\alpha}^{(0)}$ and $\hat{\beta}^{(0)}$. For example, according to Birnbaum and Saunders (1969b), we can use the mean–mean estimate of β as starting value $\hat{\beta}^{(0)}$ given by

$$\tilde{\beta} = (sr)^{1/2}, \tag{3.8}$$

where

$$s = \frac{\sum_{i=1}^{n} t_i}{n} \quad \text{and} \quad r = \frac{1}{\frac{1}{n}\sum_{i=1}^{n} \frac{1}{t_i}} \tag{3.9}$$

are the arithmetic and harmonic means of the data $t_1, t_2, \ldots, t_n$, respectively. Thus, based on the expression for $\hat{\alpha}$ given in Equation (3.5) and the starting value for $\hat{\beta}$ given in Equation (3.8), we have the starting value $\hat{\alpha}^{(0)}$. Alternatively, because β is the median of the Birnbaum–Saunders distribution, then the sample median of the data $t_1, t_2, \ldots, t_n$ can be used as starting value $\hat{\beta}^{(0)}$. Hence, again proceeding as before, we have the starting value $\hat{\alpha}^{(0)}$ for the iterative procedure presented in Equation (3.7).

Once the point estimates of α and β have been established, we can derive inference for large samples based on the properties of the maximum likelihood estimators. An aspect that facilitates this inference is the Hessian matrix. The Hessian matrix corresponds to the second derivatives of Equation (3.3) or to the first derivatives of the elements of the score vector expressed in Equation (3.4) with respect to the unknown parameters given by

$$\ddot{\ell} = \left(\frac{\partial^2 \ell(\boldsymbol{\theta})}{\partial\theta_i \partial\theta_j}\right) = \begin{pmatrix} \ddot{\ell}_{\alpha\alpha} & \ddot{\ell}_{\alpha\beta} \\ \ddot{\ell}_{\beta\alpha} & \ddot{\ell}_{\beta\beta} \end{pmatrix}, \quad i,j = 1,2, \tag{3.10}$$

where

$$\ddot{\ell}_{\alpha\alpha} = \frac{\partial^2 \ell(\alpha,\beta)}{\partial\alpha^2} = \frac{6n}{\alpha^4} - \frac{3}{\alpha^4}\sum_{i=1}^{n}\left(\frac{t_i}{\beta} + \frac{\beta}{t_i}\right) + \frac{n}{\alpha^2}, \tag{3.11}$$

$$\ddot{\ell}_{\alpha\beta} = \frac{\partial^2 \ell(\alpha,\beta)}{\partial\alpha\partial\beta} = \ddot{\ell}_{\beta\alpha} = \frac{1}{\alpha^3}\sum_{i=1}^{n}\left(\frac{1}{t_i} - \frac{t_i}{\beta^2}\right), \tag{3.12}$$

$$\ddot{\ell}_{\beta\beta} = \frac{\partial^2 \ell(\alpha,\beta)}{\partial\beta^2} = -\frac{1}{\alpha^2\beta^3}\sum_{i=1}^{n} t_i + \frac{n}{2\beta^2} - \sum_{i=1}^{n}\frac{1}{(t_i+\beta)^2}. \tag{3.13}$$

The corresponding expected Fisher information matrix can be obtained from the Hessian matrix given in Equation (3.10) by

$$\begin{aligned} \mathcal{I}(\boldsymbol{\theta}) = -\mathrm{E}\left(\frac{\partial^2 \ell(\boldsymbol{\theta})}{\partial\theta_i \partial\theta_j}\right) &= \begin{pmatrix} -\mathrm{E}(\ddot{\ell}^*_{\alpha\alpha}) & -\mathrm{E}(\ddot{\ell}^*_{\alpha\beta}) \\ -\mathrm{E}(\ddot{\ell}^*_{\beta\alpha}) & -\mathrm{E}(\ddot{\ell}^*_{\beta\beta}) \end{pmatrix} \\ &= \begin{pmatrix} \mathcal{I}_{\alpha\alpha} & \mathcal{I}_{\alpha\beta} \\ \mathcal{I}_{\beta\alpha} & \mathcal{I}_{\beta\beta} \end{pmatrix}, \quad i,j = 1,2, \end{aligned} \tag{3.14}$$

where $\ddot{\ell}^*_{\alpha\alpha}$, $\ddot{\ell}^*_{\alpha\beta}$, $\ddot{\ell}^*_{\beta\alpha}$, and $\ddot{\ell}^*_{\beta\beta}$ denote to $\ddot{\ell}_{\alpha\alpha}$, $\ddot{\ell}_{\alpha\beta}$, $\ddot{\ell}_{\beta\alpha}$, and $\ddot{\ell}_{\beta\beta}$, given, respectively, in Equations (3.11)–(3.13), evaluated at $t_i = T_i$, with $T_i \sim \mathrm{BS}(\alpha, \beta)$, for all $i = 1, \ldots, n$, such that

$$\mathrm{E}(T) = \frac{\beta}{2}\left(2+\alpha^2\right), \quad \mathrm{E}\left(\frac{1}{T}\right) = \frac{1}{2\beta}\left(2+\alpha^2\right). \tag{3.15}$$

Therefore, from Equation (3.15), we have $\mathcal{I}_{\alpha\beta} = \mathcal{I}_{\beta\alpha} = -\mathrm{E}(\ddot{\ell}^*_{\alpha\alpha}) = 0$, and then the parameters α and β are orthogonal, that is, the estimators $\hat{\alpha}$ and $\hat{\beta}$ are independent. In addition, also from Equation (3.15), we obtain

$$\mathcal{I}_{\alpha\alpha} = -\mathrm{E}(\ddot{\ell}^*_{\alpha\alpha}) = \frac{2n}{\alpha^2}, \quad \mathcal{I}_{\beta\beta} = -\mathrm{E}(\ddot{\ell}^*_{\beta\beta}) = \frac{n}{\beta^2}\left(\frac{1}{4} + \frac{1}{\alpha^2} + I(\alpha)\right),$$

with

$$I(\alpha) = 2\int_0^\infty \left(\frac{1}{1+\frac{1}{\xi(\alpha z)}} - \frac{1}{2}\right)^2 \phi(z)\,\mathrm{d}z,$$

where $\phi(\cdot)$ and $\xi(\cdot)$ are given, respectively, in Equations (2.4) and (2.6); see details in Engelhardt et al. (1981). Thus, the expected Fisher information matrix $\mathcal{I}(\boldsymbol{\theta})$ defined in Equation (3.14) is established as

$$\mathcal{I}(\boldsymbol{\theta}) = \begin{pmatrix} \frac{2n}{\alpha^2} & 0 \\ 0 & \frac{n}{\beta^2}\left(\frac{1}{4} + \frac{1}{\alpha^2} + I(\alpha)\right) \end{pmatrix}. \tag{3.16}$$

It can be verified that the corresponding regularity conditions (see Cox and Hinkley, 1974) are satisfied. Then, the estimators $\hat{\alpha}$ and $\hat{\beta}$ are consistent and have a bivariate normal joint asymptotic distribution with asymptotic means α and β, respectively, and an asymptotic covariance matrix $\boldsymbol{\Sigma}_{\hat{\theta}}$ that can be obtained from the expected Fisher information matrix given in Equation (3.16). Then, recalling $\boldsymbol{\theta} = (\alpha, \beta)^\top$, we have, as $n \to \infty$,

$$\sqrt{n}(\hat{\boldsymbol{\theta}} - \boldsymbol{\theta}) \stackrel{\cdot}{\sim} \mathrm{N}_2(\mathbf{0}_{2\times 1}, \boldsymbol{\Sigma}_{\hat{\theta}} = \mathcal{J}(\boldsymbol{\theta})^{-1}), \tag{3.17}$$

where $\mathbf{0}_{2\times 1}$ is a 2×1 vector of zeros and

$$\mathcal{J}(\boldsymbol{\theta}) = \lim_{n\to\infty} \frac{1}{n}\mathcal{I}(\boldsymbol{\theta}),$$

with $\mathcal{I}(\boldsymbol{\theta})$ being the expected Fisher information matrix given in Equation (3.16). Note that $\hat{\mathcal{I}}(\boldsymbol{\theta})^{-1}$ is a consistent estimator of the asymptotic variance–covariance matrix of $\hat{\boldsymbol{\theta}}$, $\boldsymbol{\Sigma}_{\hat{\theta}}$. In practice, one may approximate the expected Fisher information matrix by its observed version, whereas the elements of the diagonal of the inverse of this matrix can be used to approximate the corresponding standard errors; see Efron and Hinkley (1978) for details about the use of observed versus expected Fisher information matrices.

Once again note that the result given in (3.17), such as in (1.3), corresponds strictly speaking to a convergence in distribution, but we use such a notation for simplicity.

Based on the asymptotic normality, as $n \to \infty$, of the maximum likelihood estimators, denoted by $\hat{\boldsymbol{\theta}} = (\hat{\alpha}, \hat{\beta})^\top$, given in Equation (3.17), we can construct hypothesis tests and confidence intervals for $\boldsymbol{\theta}$. Note that

$$(\boldsymbol{\theta} - \hat{\boldsymbol{\theta}})^\top \boldsymbol{\Sigma}_{\hat{\theta}}^{-1} (\boldsymbol{\theta} - \hat{\boldsymbol{\theta}}) \;\dot{\sim}\; \chi^2(2)$$

is obtained from this asymptotic normality of the maximum likelihood estimators provided in Equation (3.17). Therefore, an approximate $(1-\zeta) \times 100\%$ confidence region for $\boldsymbol{\theta}$, with $0 < \zeta < 1$, is given by

$$\mathcal{R} \equiv \left\{\boldsymbol{\theta} \in \mathbb{R}^2 : (\boldsymbol{\theta} - \hat{\boldsymbol{\theta}})^\top \widehat{\boldsymbol{\Sigma}}_{\hat{\theta}}^{-1} (\boldsymbol{\theta} - \hat{\boldsymbol{\theta}}) \leq \chi^2_{1-\zeta}(2)\right\}, \tag{3.18}$$

where $\chi^2_{1-\zeta}(2)$ denotes the $(1 - \zeta) \times 100$th quantile of the central χ^2 distribution with two degrees of freedom and $\widehat{\boldsymbol{\Sigma}}$ is an estimate of $\boldsymbol{\Sigma}$.

If right censored data are present in the observed values $t_1, t_2, \ldots, t_n$ of the random sample $T_1, T_2, \ldots, T_n$ from the random variable $T \sim \text{BS}(\alpha, \beta)$, the corresponding log-likelihood function is now expressed by

$$\ell(\boldsymbol{\theta}) = \sum_{i \in D} \log\left(f_T(t_i; \boldsymbol{\theta})\right) + \sum_{i \in C} \log\left(R_T(t_i; \boldsymbol{\theta})\right), \tag{3.19}$$

where once again $\boldsymbol{\theta} = (\alpha, \beta)^\top$, $f_T(\cdot)$ and $R_T(\cdot)$ denote the probability density and reliability functions of T given in Equations (2.2) and (2.10), respectively, and D and C the uncensored and censored data sets, respectively. The log-likelihood function expressed in Equation (3.19) is also valid for type-I, type-II, and random censoring. For example, the log-likelihood function based on a right type-II censored sample (with the censored largest $n - r$ observations) from $T \sim \text{BS}(\alpha, \beta)$ is given by

$$\begin{aligned} \ell(\boldsymbol{\theta}) = & c_2 - \frac{r}{\alpha^2} - \frac{1}{2\alpha^2} \sum_{i=1}^{r} \left(\frac{t_i}{\beta} + \frac{\beta}{t_i}\right) - r\log(\alpha) \\ & - \frac{r}{2}\log(\beta) + \sum_{i=1}^{r} \log(t_i + \beta) + (n-r)\log\left(\Phi\left(-\frac{1}{\alpha}\xi\left(\frac{t}{\beta}\right)\right)\right), \end{aligned} \tag{3.20}$$

where c_2 is a constant that does not depend on $\boldsymbol{\theta}$. We recall that $\xi(\cdot)$ is defined in Equation (2.6) and $\Phi(\cdot)$ is the standard normal cumulative distribution function. Thus, in the case of Equation (3.20), the score vector $\dot{\boldsymbol{\ell}} = (\dot{\ell}_\alpha, \dot{\ell}_\beta)^\top$ has elements given by

$$\dot{\ell}_\alpha = -\frac{r}{\alpha} + \frac{1}{\alpha^3}\sum_{i=1}^{r}\left(\frac{t_i}{\beta} + \frac{\beta}{t_i} - 2\right) + \frac{2(n-r)}{\alpha}\frac{(t_{r:n} - \beta)h_T(t_{r:n}; \boldsymbol{\theta})}{(t_{r:n} + \beta)t_{r:n}},$$

$$\dot{\ell}_\beta = -\frac{r}{2\beta} + \sum_{i=1}^{r}\frac{1}{t_i + \beta} - \frac{1}{2\alpha^2}\sum_{i=1}^{r} 1\left(\frac{1}{t_i} - \frac{t_i}{\beta^2}\right) + \frac{(n-r)}{\beta}t_{r:n}h_T(t_{r:n}; \boldsymbol{\theta}), \tag{3.21}$$

where $h_T(\cdot)$ is the Birnbaum–Saunders failure rate given in Equation (2.11) and $t_{r:n}$ the observed value of the rth order statistic $T_{r:n}$ generated from $T_1, T_2, \ldots, T_n$. Thus, by solving the equations $\dot{\ell}_\alpha = 0$ and $\dot{\ell}_\beta = 0$ obtained from Equation (3.21), we have

$$\hat{\alpha} = \left(\frac{1}{r}\sum_{i=1}^{r}\left(\frac{t_i}{\hat{\beta}} + \frac{\hat{\beta}}{t_i} - 2\right) + \frac{2(n-r)\hat{\alpha}^2}{r}\frac{(t_{r:n} - \hat{\beta})h_T(t_{r:n}; \boldsymbol{\theta})}{(t_{r:n} + \hat{\beta})t_{r:n}}\right)^{1/2},$$

$$\hat{\beta} = \frac{\sum_{i=1}^{r}\frac{t_i}{2\hat{\alpha}^2}}{\sum_{i=1}^{r}\frac{1}{2\hat{\alpha}^2 t_i} - \sum_{i=1}^{r}\frac{1}{t_i + \hat{\beta}} + \frac{r}{2\hat{\beta}} - \frac{(n-r)}{\hat{\beta}}t_{r:n}h_T(t_{r:n}; \boldsymbol{\theta})}, \tag{3.22}$$

which once again requires the use of a numerical iterative procedure. In this case, the elements of the Hessian matrix are given by

$$\ddot{\ell}_{\alpha\alpha} = \frac{n}{\alpha^4} - \frac{3}{\alpha^4}\sum_{i=1}^{n}\left(\frac{t_i}{\beta} + \frac{\beta}{t_i} - 2\right) + \frac{2(n-r)t_{r:n}h_T(t_{r:n}; \boldsymbol{\theta})}{\alpha^2}\frac{(t_{r:n} - \beta)}{(t_{r:n} + \beta)}$$
$$\times\left(1 - \frac{1}{\alpha^2}\xi^2\left(\frac{t_{r:n}}{\beta}\right) - \frac{2(t_{r:n} - \beta)}{(t_{r:n} + \beta)}t_{r:n}h_T(t_{r:n}; \boldsymbol{\theta}) - t_{r:n}\right),$$

$$\ddot{\ell}_{\alpha\beta} = \frac{1}{\alpha^3}\left(\sum_{i=1}^{n}\left(\frac{1}{t_i} - \frac{t_i}{\beta^2}\right)\right)$$
$$+ \frac{(n-r)}{\alpha\beta}t_{r:n}h_T(t_{r:n})\left(\frac{1}{\alpha^2}\xi^2\left(\frac{t_{r:n}}{\beta}\right) - 1 - \frac{2(t_{r:n} - \beta)}{(t_{r:n} + \beta)}t_{r:n}h_T(t_{r:n}; \boldsymbol{\theta})\right),$$

$$\ddot{\ell}_{\beta\beta} = \frac{n}{2\beta^2} - \sum_{i=1}^{n}\frac{1}{(t_i + \beta)^2} - \frac{1}{\alpha^2\beta^3}\sum_{i=1}^{n} t_i + \frac{(n-r)}{\beta^2}t_{r:n}h_T(t_{r:n}; \boldsymbol{\theta})$$
$$\times\left(1 + \frac{(\beta - t_{r:n})}{2(\beta + t_{r:n})} - \frac{1}{2\alpha^2}\left(\frac{t_{r:n}}{\beta} - \frac{\beta}{t_{r:n}}\right) - t_{r:n}h_T(t_{r:n}; \boldsymbol{\theta})\right).$$

For more details about estimation of parameters of the Birnbaum–Saunders distribution based on censored data, the interested reader is referred to Rieck (1995), McCarter (1999), Ng et al. (2006), Wang et al. (2006), and Leiva et al. (2007). Confidence intervals and hypothesis testing for the parameter $\boldsymbol{\theta} = (\alpha, \beta)^\top$ of the Birnbaum–Saunders distribution with censored data can be conducted in a similar way as in the case of uncensored (complete) samples detailed above.

3.3 MOMENT ESTIMATION METHOD

As it is well known, moment estimators could not be unique, and, what is more, they could not always exist. This occurs when the parameters of the Birnbaum–Saunders distribution are estimated by using the standard moment method. An alternative way to estimate the parameters of the Birnbaum–Saunders distribution is by employing the modified moment method, which provides estimators that always exist and are unique; see Ng et al. (2003). Next, we present the modified moment estimators of the parameters α and β of the Birnbaum–Saunders distribution and their asymptotic distributions, which can be used to construct confidence intervals and hypothesis testing for these parameters. The modified moment estimation method is a particular case of the generalized method of moment estimation. The interested reader is referred to Mátyás (1999) for seeing details about the generalized moment estimation method and to Santos-Neto et al. (2014) for the application of this method to a reparameterized version of the Birnbaum–Saunders distribution proposed by Santos-Neto et al. (2012); see also Leiva et al. (2014c).

Let $T_1, T_2, \ldots, T_n$ be a sample of size n from $T \sim \mathrm{BS}(\alpha, \beta)$. Then, the modified moment estimates of α and β, denoted by $\tilde{\alpha}$ and $\tilde{\beta}$, respectively, are given by

$$\tilde{\alpha} = \left(2\left(\sqrt{\frac{s}{r}} - 1\right)\right)^{1/2} \quad \text{and} \quad \tilde{\beta} = \sqrt{sr}. \tag{3.23}$$

We recall s and r are the arithmetic and geometric means, respectively, defined in Equation (3.9). Note that the modified moment estimate of β coincides with the mean–mean estimate proposed by Birnbaum and Saunders (1969b) and given in Equation (3.8).

The asymptotic joint distribution of the modified moment estimators $\tilde{\alpha}$ and $\tilde{\beta}$ is bivariate normal and given by

$$\sqrt{n}\left(\begin{pmatrix}\tilde{\alpha}\\ \tilde{\beta}\end{pmatrix} - \begin{pmatrix}\alpha\\ \beta\end{pmatrix}\right) \dot{\sim} \mathrm{N}_2(\mathbf{0}_{2\times 1}, \mathbf{\Sigma}_{\tilde{\theta}}),$$

as $n \to \infty$, where $\mathbf{0}_{2\times 1}$ is a 2×1 vector of zeros and

$$\mathbf{\Sigma}_{\tilde{\theta}} = \begin{pmatrix}\frac{\alpha^2}{2} & 0\\ 0 & (\alpha\beta)^2\frac{(4+3\alpha^2)}{(2+\alpha^2)^2}\end{pmatrix}$$

is the asymptotic variance–covariance matrix of $\tilde{\boldsymbol{\theta}}$. Thus, the asymptotic marginal distributions of $\tilde{\alpha}$ and $\tilde{\beta}$, as $n \to \infty$, are given by

$$\sqrt{n}\,(\tilde{\alpha} - \alpha) \dot{\sim} \mathrm{N}\left(0, \frac{\alpha^2}{2}\right), \tag{3.24}$$

$$\sqrt{n}(\tilde{\beta} - \beta) \dot{\sim} \mathrm{N}\left(0, (\alpha\beta)^2\frac{(4+3\alpha^2)}{(2+\alpha^2)^2}\right),$$

respectively. From the asymptotic results given in Equation (3.24), we can construct confidence intervals and hypothesis testing for the parameters α and β of the Birnbaum–Saunders distribution.

3.4 GRAPHICAL ESTIMATION METHOD

Chang and Tang (1994a) developed a simple graphical method analogous to the probability versus probability plots for the Birnbaum–Saunders distribution. This graphical method is useful as a visual goodness-of-fit tool and can also be used as an estimation method for the parameters of the Birnbaum–Saunders distribution, or at least for finding initial values for the iterative procedure needed in the maximum likelihood method. In the graphical method, the data are transformed allowing pairs of values to be plotted. By using a simple linear regression method, the slope and the intercept of the line are estimated. This line is used for goodness of fit, such as occurs with probability plots. Thus, if we consider the cumulative distribution function $F_T(\cdot)$ of a random variable T with Birnbaum–Saunders distribution given in Equation (1.6), we have

$$t = \alpha\sqrt{\beta}\sqrt{t}\Phi^{-1}(F_T(t)) + \beta. \tag{3.25}$$

However, from Equation (3.25), it is difficult to derive a linear function over t, which is fundamental for probability plotting. Chang and Tang (1994a) considered

$$p = \sqrt{t}\Phi^{-1}(F_T(t)) \tag{3.26}$$

and obtained the linear function

$$y \approx mx + b, \tag{3.27}$$

where the x-axis is $x = p$, with p being defined in Equation (3.26) and the y-axis is $y = t$, whereas the intercept and the slope are, respectively, given by

$$m = \alpha\sqrt{\beta} \quad \text{and} \quad b = \beta. \tag{3.28}$$

Let $T_1, T_2, \ldots, T_n$ be a sample of size n from $T \sim \mathrm{BS}(\alpha, \beta)$ and $t_1, t_2, \ldots, t_n$ their observations. Then, plotting t_i versus $\bar{p}_i$, where

$$\bar{p}_i = \sqrt{t_i}\Phi^{-1}(\bar{F}_T(t_i)),$$

the result is approximately a straight line if the data come from a Birnbaum–Saunders distribution, with

$$\bar{F}_T(t_i) = \frac{i - 0.3}{n + 0.4}, \quad i = 1, 2, \ldots, n,$$

being the mean rank; see Chang and Tang (1994a). Goodness of fit for the Birnbaum–Saunders distribution, based on t_i versus $\bar{p}_i$, can be visually and analytically studied using the coefficient of determination (R^2).

By using the least square method, we can estimate the slope m and intercept b of the linear function given in Equation (3.27) as $\bar{m}$ and $\bar{b}$, respectively. Then, the least square estimates of the parameters α and β are

$$\bar{\beta} = \bar{b} \quad \text{and} \quad \bar{\alpha} = \frac{\bar{m}}{\sqrt{\bar{b}}}.$$

It can be observed that the parameter α of the Birnbaum–Saunders distribution is an increasing function over $m = \alpha\sqrt{\beta}$ and decreasing over $b = \beta$. In this way, using a result from the regression analysis, it is easy to find approximate $(1 - \zeta_1 - \zeta_2) \times 100\%$ confidence intervals for a and b and to test hypotheses for these parameters. Thus, these confidence intervals can be used to obtain approximate confidence intervals for α and β given, respectively, by

$$(\bar{\alpha}_l, \bar{\alpha}_u) = \left(\frac{\bar{m}_l}{\sqrt{\bar{b}_l}}, \frac{\bar{m}_u}{\sqrt{\bar{b}_u}}\right) \quad \text{and} \quad (\bar{\beta}_l, \bar{\beta}_u) = \left(\bar{b}_l, \bar{b}_u\right), \tag{3.29}$$

where $(\bar{m}_l, \bar{m}_u)$ and $(\bar{b}_l, \bar{b}_u)$ are approximate confidence intervals for m and b with confidence levels $(1 - \zeta_1) \times 100\%$ and $(1 - \zeta_2) \times 100\%$, respectively.

CHAPTER 4

Modeling Based on the Birnbaum–Saunders Distribution

Abstract

In this chapter, regression models and their diagnostics based on the Birnbaum–Saunders distribution with uncensored and censored data are presented. We detail the maximum likelihood method for estimating the regression parameters and the associated asymptotic inference. In addition, we derive influence diagnostics and residual analysis for these models. The local influence method is considered using weight-case, response variable and explanatory variable perturbation schemes. The generalized leverage method is also studied. Deviance component and martingale-type residuals are analyzed.

Keywords: asymptotic inference, censored data, failure rate, generalized leverage, least square method, local influence, log-Birnbaum–Saunders distribution, maximum likelihood method, probability density function, regression, reliability function, residuals, sinh-normal distribution, standard normal distribution.

4.1 INTRODUCTION

Statistical modeling based on the Birnbaum–Saunders distribution has been considered by Rieck and Nedelman (1991), Owen and Padgett (1999), and Tsionas (2001), whereas Galea et al. (2004), Leiva et al. (2007), and Xie and Wei (2007) studied aspects related to influence diagnostics in log-Birnbaum–Saunders regression models with noncensored and censored data.

A way to carry out a modeling based on the Birnbaum–Saunders distribution is by means of the sinh-normal distribution proposed by Rieck and Nedelman (1991). The sinh-normal distribution has the log-Birnbaum–Saunders distribution mentioned in Chapter 3 as a particular case. The sinh-normal distribution is symmetrical, presents greater and smaller degrees of kurtosis than the normal distribution, and admits unimodality and bimodality.

The Birnbaum–Saunders Distribution. http://dx.doi.org/10.1016/B978-0-12-803769-0.00004-2

4.2 THE LOG-BIRNBAUM–SAUNDERS DISTRIBUTION

The sinh-normal distribution is obtained from the transformation

$$Y = \mu + \sigma \operatorname{arcsinh}\left(\frac{\alpha Z}{2}\right),$$

where $Z \sim \mathrm{N}(0,1)$, $\alpha > 0$ is a shape parameter, $\mu \in \mathbb{R}$ is a location parameter, and $\sigma > 0$ is a scale parameter. In this case, the notation $Y \sim \mathrm{SHN}(\alpha, \mu, \sigma)$ is used. The probability density function of Y is given by

$$f_Y(y; \alpha, \mu, \sigma) = \phi\left(\frac{2}{\alpha}\sinh\left(\frac{y-\mu}{\sigma}\right)\right)\frac{2\cosh\left((y-\mu)/\sigma\right)}{\alpha\,\sigma}, \quad y \in \mathbb{R}. \tag{4.1}$$

The cumulative distribution function of Y is expressed by

$$F_Y(y; \alpha, \mu, \sigma) = \Phi\left(\frac{2}{\alpha}\sinh\left(\frac{y-\mu}{\sigma}\right)\right), \quad y \in \mathbb{R}. \tag{4.2}$$

The quantile function of Y is defined as

$$y(q; \alpha, \mu, \sigma) = F_Y^{-1}(q; \alpha, \mu, \sigma) = \mu + \sigma \operatorname{arcsinh}\left(\frac{\alpha z(q)}{2}\right), \quad 0 < q \leq 1,$$

where $z(q)$ is the $q \times 100$th quantile of $Z \sim \mathrm{N}(0,1)$ and $F_Y^{-1}(\cdot)$ is the inverse function of $F_Y(\cdot)$. The mean and variance of Y can be obtained using its moment generating function given by

$$m(r) = \exp(r\mu)\left(\frac{K_{r\sigma+1/2}(1/\delta^2) + K_{r\sigma-1/2}(1/\delta^2)}{2K_{1/2}(1/\delta^2)}\right), \tag{4.3}$$

where $\mathrm{E}(Y) = \mu$ and $K_\lambda(\cdot)$ is the modified Bessel function of the third kind detailed in Chapter 3.

The reliability function and failure rate of $Y \sim \mathrm{SHN}(\alpha, \mu, \sigma)$ are given by

$$R_Y(y; \alpha, \mu, \sigma) = \Phi\left(-\frac{2}{\alpha}\sinh\left(\frac{y-\mu}{\sigma}\right)\right), \quad y \in \mathbb{R}, \tag{4.4}$$

$$h_Y(y; \alpha, \mu, \sigma) = \frac{2\,\phi\left(\frac{2}{\alpha}\sinh\left(\frac{y-\mu}{\sigma}\right)\right)\cosh\left(\frac{y-\mu}{\sigma}\right)}{\alpha\,\sigma\,\Phi\left(-\frac{2}{\alpha}\sinh\left(\frac{y-\mu}{\sigma}\right)\right)}, \quad y \in \mathbb{R}, \tag{4.5}$$

respectively. Rieck (1989) noted that a sinh-normal distribution with parameters α, μ, and σ is:

(E1) Symmetrical around its mean μ.
(E2) Unimodal for $\alpha < 2$ and its kurtosis is smaller than that normal kurtosis.

(E3) Platykurtic for $\alpha = 2$.

(E4) Bimodal for $\alpha > 2$ and, as α increases, the distribution begins to emphasize its bimodality, with modes that are more separated, and its kurtosis is greater than the kurtosis of the normal distribution.

In addition, if $Y \sim \text{SHN}(\alpha, \mu, \sigma)$, then, as α approaches zero,

$$\frac{2(Y - \mu)}{\alpha\,\sigma} \dot{\sim} \text{N}(0, 1).$$

Thus, as mentioned, the parameter α modifies the shape of the sinh-normal distribution, because as α increases, the kurtosis also increases. The parameter μ, however, modifies the location, while the parameter σ modifies the scale of this distribution, all which can be verified in Figure 4.1.

Rieck and Nedelman (1991) proved that if $T \sim \text{BS}(\alpha, \beta)$, then $Y = \log(T) \sim \text{SHN}(\alpha, \mu, \sigma = 2)$, where $\mu = \log(\beta)$. For this reason, the sinh-normal distribution is also known as log-Birnbaum–Saunders distribution. Thus, based on Equation (4.3), $Y = \log(T)$ say, and the relationship

$$M_Y(r) = \text{E}(\exp(r\,Y)) = \text{E}(T^r), \tag{4.6}$$

the moments of any order of the Birnbaum–Saunders distribution can be computed; see Rieck (1999). In addition, estimation of parameters of the Birnbaum–Saunders distribution and generation of random numbers from this distribution can be more efficiently obtained from the log-Birnbaum–Saunders distribution. For more details, see Chapter 3.

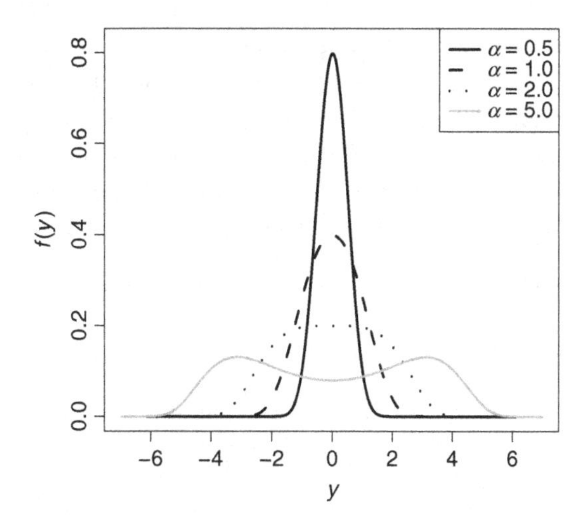

Figure 4.1 Plots of the log-Birnbaum–Saunders probability density function for the indicated value of α with $\mu = 0$.

4.3 REGRESSION MODELS

Next, regression models for the Birnbaum–Saunders distribution based on logarithmic transformations of the observations, including uncensored and censored data, are introduced. These models can also be useful for data that follow the sinh-normal distribution or some of its generalizations without having to transform them; see Leiva et al. (2007), Barros et al. (2008), and Paula et al. (2012).

The linear models presented here can be used in a wide variety of applications; see Rieck and Nedelman (1991). The corresponding maximum likelihood estimators of the model coefficients do not have a closed form so that their exact distribution cannot be found. Therefore, we rely on large sample asymptotic results of normal theory procedures to make inferences concerning to the Birnbaum–Saunders regression model coefficients.

As mentioned, the sinh-normal distribution is symmetrical and, for small α, it converges to a normal distribution. In addition, because the shape parameter α is assumed to be constant, the Birnbaum–Saunders regression models exhibit homoscedasticity. Normal theory methods for means are known to be robust against departures from the normality, even if symmetry and homoscedasticity do not hold; see Kendall and Stuart (1974) and Díaz-García and Leiva (2003). Hence, t or $\mathcal{F}$ tests may also be used to compare location parameters in two-or-multi-samples. Interval confidence for the regression coefficients may be constructed with least square estimates using the Student t distribution and an estimate of the variance of the estimators of these coefficients. For large sample sizes, likelihood methods may be applied. Specifically, confidence intervals of such coefficients may be obtained using the maximum likelihood estimates, the normal distribution and the Fisher information matrix. Achcar and Espinosa (1992) and Tsionas (2001) discussed Bayesian linear regression models when the errors follow a Birnbaum–Saunders distribution. Note that no important differences between the maximum likelihood and Bayesian regression parameter estimates were detected.

Consider the Birnbaum–Saunders regression model

$$T_i = \beta_i \varphi_i = \exp(\mu_i)\varphi_i = \exp(\boldsymbol{x}_i^\top \boldsymbol{\eta})\, \varphi_i, \quad i = 1, 2, \ldots, n, \tag{4.7}$$

where T_i and $\beta_i = \exp(\mu_i)$ are the response variable and median for the case i, respectively, $\boldsymbol{\eta} = (\eta_0, \eta_1, \ldots, \eta_p)^\top$ is a $(p+1) \times 1$ vector of unknown parameters to be estimated, $\boldsymbol{x}_i^\top = (1, x_{i1}, \ldots, x_{ip})$ contains values of p

explanatory variables and $\varphi_i \sim \text{BS}(\alpha, 1)$ is the model error, for $i = 1, \ldots, n$. Note that, using Property (D1) of the Birnbaum–Saunders distribution, $T_i \sim \text{BS}(\alpha, \beta_i)$. Now, by applying logarithm in Equation (4.7), we obtain

$$Y_i = \mu_i + \varepsilon_i = \boldsymbol{x}_i^\top \boldsymbol{\eta} + \varepsilon_i, \quad i = 1, 2, \ldots, n, \tag{4.8}$$

where $Y_i = \log(T_i)$ is the log-response variable for the case i, $\boldsymbol{\eta}$ and $\boldsymbol{x}_i$ are as given in Equation (4.7) and $\varepsilon_i = \log(\varphi_i) \sim \text{log-BS}(\alpha, 0)$ is the error term of the model, for $i = 1, \ldots, n$.

Consider the Birnbaum–Saunders regression model given in Equation (4.7) such that $T_1, T_2, \ldots, T_n$ is a sample of size n from $T \sim \text{BS}(\alpha, \beta_i)$. Then, $Y_1 = \log(T_1), \log(T_2), \ldots, Y_n = \log(T_n)$ can be considered as a sample of size n from $Y \sim \text{log-BS}(\alpha, \mu_i = \log(\beta_i))$, with $y_1, \ldots, y_n$ being their observations. Hence, the log-likelihood function for $\boldsymbol{\theta} = (\alpha, \boldsymbol{\eta}^\top)^\top$ based on $y_1, \ldots, y_n$ is given by

$$\begin{aligned}\ell(\boldsymbol{\theta}) &= \sum_{i=1}^{n} \log\left(\frac{1}{\alpha\sqrt{2\pi}} \cosh\left(\frac{y_i - \mu_i}{2}\right) \exp\left(-\frac{2}{\alpha^2}\left(\sinh\left(\frac{y_i - \mu_i}{2}\right)\right)^2\right)\right) \\ &= c_3 + \sum_{i=1}^{n} \log(\xi_{i1}) - \frac{1}{2}\sum_{i=1}^{n} \xi_{i2}^2,\end{aligned} \tag{4.9}$$

where c_3 is a constant that does not depend on $\boldsymbol{\theta}$,

$$\xi_{i1} = \frac{2}{\alpha} \cosh\left(\frac{y_i - \mu_i}{2}\right), \quad \xi_{i2} = \frac{2}{\alpha} \sinh\left(\frac{y_i - \mu_i}{2}\right), \tag{4.10}$$

and $\mu_i = \boldsymbol{x}_i^\top \boldsymbol{\eta}$, for $i = 1, 2, \ldots, n$.

The maximum likelihood estimates of the shape parameter and regression coefficients are solutions of the equations

$$\dot{\ell}_\alpha = 0, \quad \dot{\ell}_{\eta_j} = 0, \; j = 0, 1, \ldots, p, \tag{4.11}$$

where $\dot{\ell}_\alpha$ and $\dot{\ell}_{\eta_j}$ are the first derivatives of the likelihood function given in Equation (4.9) constituting the corresponding score vector defined by

$$\dot{\boldsymbol{\ell}} = \begin{pmatrix} \frac{\partial \ell(\boldsymbol{\theta})}{\partial \alpha} \\ \frac{\partial \ell(\boldsymbol{\theta})}{\partial \boldsymbol{\eta}} \end{pmatrix} = \begin{pmatrix} \dot{\ell}_\alpha \\ \dot{\boldsymbol{\ell}}_\eta \end{pmatrix}, \tag{4.12}$$

with

$$\dot{\ell}_\alpha = \frac{1}{\alpha}\sum_{i=1}^{n}\left(\xi_{i2}^2 - 1\right), \quad \dot{\boldsymbol{\ell}}_\eta = (\dot{\ell}_{\eta_0}, \dot{\ell}_{\eta_1}, \ldots, \dot{\ell}_{\eta_p})^\top,$$

and

$$\dot{\ell}_{\eta_j} = \sum_{i=1}^{n} \left(\frac{x_{ij}}{\alpha^2} \sinh(y_i - \mu_i) - \frac{x_{ij}}{2} \tanh\left(\frac{y_i - \mu_i}{2}\right)\right), \quad j = 0, 1, \ldots, p.$$

By solving the log-likelihood equation $\dot{\ell}_\alpha = 0$ given in Equation (4.11) on α and evaluating it at $\hat{\alpha}$, the maximum likelihood estimate of the shape parameter is obtained as

$$\hat{\alpha} = \sqrt{\frac{4}{n} \sum_{i=1}^{n} \sinh^2\left(\frac{y_i - \boldsymbol{x}_i^\top \hat{\boldsymbol{\eta}}}{2}\right)},$$

where $\hat{\boldsymbol{\eta}}$ is the maximum likelihood estimate of $\boldsymbol{\eta}$. However, equations $\dot{\ell}_{\eta_j} = 0$, for $j = 0, 1, \ldots, p$, also given in Equation (4.11), do not present analytical solutions. Then, the use of iterative methods for solving nonlinear optimization problems is needed. For instance, the Broyden–Fletcher–Goldfarb–Shanno (BFGS) quasi-Newton method can be used; see Nocedal and Wright (1999), Lange (2001), and Press et al. (1992). The BFGS method is available in the `R` software by the functions `optim` and `optimx`; see www.R-project.org and R-Team (2015).

Note that the bimodality of the log-Birnbaum–Saunders distribution when $\alpha > 2$ can cause multiple maxima in the corresponding log-likelihood function. However, experience with fatigue data indicates that the case of $\alpha > 2$ is unusual in practice. Also, it has been shown that, if $\alpha \leq 2$, the maximum likelihood estimate of $\boldsymbol{\eta}$ is unique if $\boldsymbol{X} = (\boldsymbol{x}_1, \ldots, \boldsymbol{x}_n)^\top$ has rank p. Rieck and Nedelman (1991) found that the least square estimator of $\boldsymbol{\eta}$ is less efficient than its maximum likelihood estimator. As it is well known, the least square estimate of $\boldsymbol{\eta}$ is given by

$$\bar{\boldsymbol{\eta}} = (\boldsymbol{X}^\top \boldsymbol{X})^{-1} \boldsymbol{X}^\top \boldsymbol{y},$$

where $\boldsymbol{y}$ is an $n \times 1$ vector of observations and $\boldsymbol{X}$ the design matrix of the model.

Similarly to Chapter 3, inference in large samples for $\boldsymbol{\theta}$ can again be based on the asymptotic normality of the estimator $\hat{\boldsymbol{\theta}} = (\hat{\alpha}, \hat{\boldsymbol{\eta}}^\top)^\top$, as $n \to \infty$, which is given by

$$\hat{\boldsymbol{\theta}} \overset{\cdot}{\sim} \mathrm{N}_{p+2}(\boldsymbol{\theta}, \boldsymbol{\Sigma}_{\hat{\theta}}), \tag{4.13}$$

where $\boldsymbol{\Sigma}_{\hat{\theta}}$ is the $(p + 2) \times (p + 2)$ asymptotic variance–covariance matrix de $\hat{\boldsymbol{\theta}}$; see details about this result in (3.17) of Chapter 3. Note that $\boldsymbol{\Sigma}_{\hat{\theta}}$ can be approximated by $-\ddot{\boldsymbol{\ell}}^{-1}$, where $-\ddot{\boldsymbol{\ell}}$ is the $(p + 2) \times (p + 2)$ observed information matrix obtained from the Hessian matrix with second derivatives given by

$$\ddot{\boldsymbol{\ell}} = \begin{pmatrix} \frac{\partial^2 \ell(\boldsymbol{\theta})}{\partial \alpha^2} & \frac{\partial^2 \ell(\boldsymbol{\theta})}{\partial \alpha \partial \boldsymbol{\eta}} \\ \frac{\partial^2 \ell(\boldsymbol{\theta})}{\partial \boldsymbol{\eta} \partial \alpha} & \frac{\partial^2 \ell(\boldsymbol{\theta})}{\partial \boldsymbol{\eta} \partial \boldsymbol{\eta}^\top} \end{pmatrix} = \begin{pmatrix} \ddot{\ell}_{\alpha\alpha} & \ddot{\boldsymbol{\ell}}_{\alpha\eta} \\ \ddot{\boldsymbol{\ell}}_{\eta\alpha} & \ddot{\boldsymbol{\ell}}_{\eta\eta} \end{pmatrix} = \begin{pmatrix} \sum_{i=1}^{n} \frac{(1-3\xi_{i2}^2)}{\alpha^2} & \boldsymbol{k}^\top \boldsymbol{X} \\ \boldsymbol{X}^\top \boldsymbol{k} & \boldsymbol{X}^\top \boldsymbol{V} \boldsymbol{X} \end{pmatrix}, \tag{4.14}$$

where $\boldsymbol{V} = \text{diag}\{v_1(\boldsymbol{\theta}), \ldots, v_n(\boldsymbol{\theta})\}$ and $\boldsymbol{k} = (k_1, \ldots, k_n)^\top$, with

$$v_i(\boldsymbol{\theta}) = \frac{1}{4}\left(\text{sech}\left(\frac{y_i - \mu_i}{2}\right)\right)^2 - \frac{1}{\alpha^2}\cosh(y_i - \mu_i), \tag{4.15}$$

$$k_i = -\frac{2}{\alpha^3}\sinh(y_i - \mu_i), \quad i = 1, 2, \ldots, n.$$

Thus, as it is well known, based on the result given in Equation (4.13), an approximate $(1-\varrho) \times 100\%$ confidence region for the parameter $\boldsymbol{\theta}$ can be obtained from

$$\mathcal{R} \equiv \left\{\boldsymbol{\theta} \in \mathbb{R}^{p+2} : (\boldsymbol{\theta} - \hat{\boldsymbol{\theta}})^\top \widehat{\boldsymbol{\Sigma}}_{\hat{\theta}}^{-1}(\boldsymbol{\theta} - \hat{\boldsymbol{\theta}}) \leq \chi^2_{1-\varrho}(p+2)\right\},$$

where $\chi^2_{1-\varrho}(p+2)$ denotes the $(1-\varrho) \times 100$th quantile of the central chi-squared distribution with $p+2$ degrees of freedom and $\widehat{\boldsymbol{\Sigma}}$ is an estimate of $\boldsymbol{\Sigma}$, which can be obtained from the observed information matrix based on Equation (4.14) evaluated at $\hat{\boldsymbol{\theta}}$.

Similarly to Equation (4.8), once again consider Birnbaum–Saunders log-linear regression models given by

$$Y_i = \boldsymbol{x}_i^\top \boldsymbol{\eta} + \varepsilon_i, \quad i = 1, \ldots, n, \tag{4.16}$$

where Y_i is the censored or uncensored log-response variable for the case i, $\boldsymbol{\eta} = (\eta_0, \eta_1, \ldots, \eta_p)^\top$ is a vector of unknown parameters to be estimated, $\boldsymbol{x}_i^\top = (1, x_{i1}, \ldots, x_{ip})^\top$ contains values of explanatory variables, and $\varepsilon_i \sim \text{log-BS}(\alpha, 0)$ is the error term of the model, for $i = 1, \ldots, n$. We assume noninformative censoring and that the uncensored and censored log-response variables are independent.

The log-likelihood function of the model defined in Equation (4.16) for $\boldsymbol{\theta} = (\alpha, \boldsymbol{\eta}^\top)^\top$ is given by

$$\ell(\boldsymbol{\theta}) = c_4 + \sum_{i \in D} \ell_i(\boldsymbol{\theta}) + \sum_{i \in C} \ell_i^{(c)}(\boldsymbol{\theta}),$$

where D and C denote the sets of cases for which y_i is the uncensored and censored log-response, respectively, c_4 is a constant that does not depend on

$\boldsymbol{\theta}$, $\ell_i(\boldsymbol{\theta}) = \log(f_Y(y_i;\boldsymbol{\theta}))$ and $\ell_i^{(c)}(\boldsymbol{\theta}) = \log(R_Y(y_i;\boldsymbol{\theta}))$, with $f_Y(\cdot)$ and $R_Y(\cdot)$ being the probability density and reliability functions of the log-Birnbaum–Saunders distribution given in Equations (4.1) and (4.4), respectively. Then, the log-likelihood function for $\boldsymbol{\theta}$ is expressed by

$$\ell(\boldsymbol{\theta}) = c_4 + \sum_{i\in D}\left(\log(\xi_{i1}) - \frac{\xi_{i2}^2}{2}\right) + \sum_{i\in C}\log\left(\Phi(-\xi_{i2})\right), \qquad (4.17)$$

where ξ_{i1}, ξ_{i2}, and μ_i, for $i = 1, 2, \ldots, n$, are given in Equation (4.10).

To obtain the maximum likelihood estimate of $\boldsymbol{\theta}$, the log-likelihood equations $\dot{\ell}_\alpha = 0$ and $\dot{\ell}_{\eta_j}(\boldsymbol{\theta}) = 0$, for $j = 0, 1, \ldots, p$, must be solved, where

$$\dot{\ell}_\alpha = \frac{1}{\alpha}\left(\sum_{i\in D}\left(\xi_{i2}^2 - 1\right) + \sum_{i\in C}\xi_{i2}\, h_Y(\xi_{i2};\boldsymbol{\theta})\right)$$

$$\dot{\ell}_{\eta_j} = \sum_{i\in D}\left(\frac{x_{ij}}{\alpha^2}\sinh(y_i - \mu_i) - \frac{x_{ij}}{2}\tanh\left(\frac{y_i - \mu_i}{2}\right)\right) + \sum_{i\in C}\frac{x_{ij}\xi_{i1}h_Y(\xi_{i2};\boldsymbol{\theta})}{2},$$

with $h_Y(\cdot)$ being the failure rate of the log-Birnbaum–Saunders distribution given in Equation (4.5). Because once again the log-likelihood equations do not present analytical solutions, we can use the BFGS method and the maximum likelihood estimates of $\boldsymbol{\theta}$ in the uncensored case as starting values; see Leiva et al. (2007).

Such as in the uncensored case, asymptotic inference for the parameter vector $\boldsymbol{\theta}$ is considered. The asymptotic variance–covariance matrix of $\hat{\boldsymbol{\theta}}$ can be approximated by $-\ddot{\boldsymbol{\ell}}^{-1}$, which is obtained in this case from

$$\ddot{\boldsymbol{\ell}} = \begin{pmatrix} \ddot{\ell}_{\alpha\alpha} & \ddot{\ell}_{\alpha\eta} \\ \ddot{\ell}_{\eta\alpha} & \ddot{\ell}_{\eta\eta} \end{pmatrix} = \begin{pmatrix} \mathrm{tr}(\boldsymbol{G}) & \boldsymbol{k}^\top\boldsymbol{X} \\ \boldsymbol{X}^\top\boldsymbol{k} & \boldsymbol{X}^\top\boldsymbol{V}\boldsymbol{X} \end{pmatrix}, \qquad (4.18)$$

where $\boldsymbol{V} = \mathrm{diag}\{v_1(\boldsymbol{\theta}), v_2(\boldsymbol{\theta}) \ldots, v_n(\boldsymbol{\theta})\}$, $\boldsymbol{k} = (k_1(\boldsymbol{\theta}), \ldots, k_n(\boldsymbol{\theta}))^\top$, and $\boldsymbol{G} = \mathrm{diag}\{g_1(\boldsymbol{\theta}), \ldots, g_n(\boldsymbol{\theta})\}$, with

$$v_i(\boldsymbol{\theta}) = \begin{cases} \frac{1}{4}\left(\mathrm{sech}\left(\frac{y_i-\mu_i}{2}\right)\right)^2 - \frac{1}{\alpha^2}\cosh(y_i - \mu_i), & i \in D; \\ -\frac{1}{4}\xi_{i2}\, h_Y(\xi_{i2};\boldsymbol{\theta}) - \frac{1}{4}\xi_{i1}^2 h'_Y(\xi_{i2};\boldsymbol{\theta}), & i \in C; \end{cases} \qquad (4.19)$$

$$k_i(\boldsymbol{\theta}) = \begin{cases} -\frac{2}{\alpha^3}\sinh(y_i - \mu_i), & i \in D; \\ -\frac{1}{2\alpha}\xi_{i1}\, h_Y(\xi_{i2};\boldsymbol{\theta}) - \frac{1}{\alpha^3}\sinh(y_i - \mu_i)h'_Y(\xi_{i2};\boldsymbol{\theta}), & i \in C; \end{cases} \qquad (4.20)$$

$$g_i(\boldsymbol{\theta}) = \begin{cases} \frac{1}{\alpha^2} - \frac{3\xi_{i2}^2}{\alpha^2}, & i \in D; \\ -\frac{2}{\alpha^2}\xi_{i2}\, h_Y(\xi_{i2};\boldsymbol{\theta}) - \frac{1}{\alpha^2}\xi_{i2}^2 h'_Y(\xi_{i2};\boldsymbol{\theta}), & i \in C; \end{cases} \tag{4.21}$$

where $h'_Y(\cdot)$ is the derivative of $h_Y(\cdot)$, the log-Birnbaum–Saunders failure rate, and ξ_{i1}, ξ_{i2} are given in Equation (4.17).

4.4 DIAGNOSTIC METHODS

Local influence

To assess the sensitivity of the maximum likelihood estimates to atypical data, an analysis of normal curvatures of local influence developed by Cook (1986) may be carried out for some common perturbation schemes. Diagnostic methods for normal linear regression models have been largely investigated in the statistical literature; see, for example, Belsley et al. (1980), Cook and Weisberg (1982), and Chatterjee and Hadi (1988). The majority of the works on diagnostics have given emphasis in studying the effect of eliminating observations on the results from the fitted model, particularly on the parameter estimates. Alternatively, Cook (1986) proposed a method named local influence to assess the effect of small perturbations in the model and/or data on the parameter estimates. Several authors have extended the local influence method to various regression models; see, for example, Lawrance (1988), Escobar and Meeker (1992), Paula (1993), Galea et al. (1997, 2000), Ortega et al. (2003), Díaz-García et al. (2003), Leiva et al. (2007, 2016), Barros et al. (2008), and Liu et al. (2015).

Let $\ell(\boldsymbol{\theta})$ denote the corresponding log-likelihood function. The perturbations are made on $\ell(\boldsymbol{\theta})$, such that it takes the form $\ell(\boldsymbol{\theta}|\boldsymbol{\omega})$, where $\boldsymbol{\omega}$ is a $q \times 1$ perturbation vector restricted to some open subset $\Omega \subset \mathbb{R}^q$ and $\boldsymbol{\omega}_0 = (1, \ldots, 1)^\top$ is the nonperturbation vector, such that $\ell(\boldsymbol{\theta}|\boldsymbol{\omega}_0) = \ell(\boldsymbol{\theta})$. To assess the influence of the perturbations on the maximum likelihood estimate of $\boldsymbol{\theta}$, $\hat{\boldsymbol{\theta}}$ say, we consider the likelihood displacement given by

$$\mathrm{LD}(\boldsymbol{\omega}) = 2\left(\ell(\hat{\boldsymbol{\theta}}) - \ell(\hat{\boldsymbol{\theta}}_\omega)\right), \tag{4.22}$$

where $\hat{\boldsymbol{\theta}}_\omega$ denotes the maximum likelihood estimate of $\boldsymbol{\theta}$ under the model $\ell(\boldsymbol{\theta}|\boldsymbol{\omega})$.

The idea of local influence proposed by Cook (1986) consists of studying the behavior of $\mathrm{LD}(\boldsymbol{\omega})$ around $\boldsymbol{\omega}_0$. The procedure selects a direction vector $\boldsymbol{l}$, such that $\|\boldsymbol{l}\| = 1$, and then considers the plot of $\mathrm{LD}(\boldsymbol{\omega}_0 + a\,\boldsymbol{l})$ against a, for $a \in \mathbb{R}$. This plot is called the lifted line. Note that, because $\text{LD}(\boldsymbol{\omega}_0) = 0$, $\text{LD}(\boldsymbol{\omega}_0 + a\boldsymbol{l})$ has a local minimum at $a = 0$. Each lifted line can be characterized by considering the normal curvature $C_l(\boldsymbol{\theta})$ around $a = 0$. This curvature is interpreted as the inverse of the radius of the circle which provides the best fit at $a = 0$. Cook (1986) suggested considering the direction vector $\boldsymbol{l}_{\max}$ corresponding to the largest curvature $C_{l_{\max}}(\boldsymbol{\theta})$. Therefore, the index plot of $\boldsymbol{l}_{\max}$ may reveal those observations that under a small perturbation exercise a considerable influence on $\text{LD}(\boldsymbol{\omega})$. Cook (1986) showed that the normal curvature at the direction vector $\boldsymbol{l}$ takes the form

$$C_l(\boldsymbol{\theta}) = 2\left|\boldsymbol{l}^\top \boldsymbol{\Delta}^\top \ddot{\boldsymbol{\ell}}^{-1} \boldsymbol{\Delta}\boldsymbol{l}\right|, \tag{4.23}$$

where $\ddot{\boldsymbol{\ell}}$ is the Hessian matrix associated with the corresponding model and $\boldsymbol{\Delta}$ is a $(p+2) \times n$ perturbation matrix with elements

$$\Delta_{ji} = \frac{\partial^2 \ell(\boldsymbol{\theta}|\boldsymbol{\omega})}{\partial\theta_j \partial\omega_i}, \quad i = 1, 2, \ldots, n, \; j = 1, 2, \ldots, p+2. \tag{4.24}$$

Note that the perturbation matrix $\boldsymbol{\Delta}$ with elements given in Equation (4.24) must be evaluated at $\boldsymbol{\theta} = \hat{\boldsymbol{\theta}}$ and $\boldsymbol{\omega} = \boldsymbol{\omega}_0$. Then, $C_{l_{\max}}$ is the largest eigenvalue of the matrix

$$\boldsymbol{B} = \boldsymbol{\Delta}^\top \ddot{\boldsymbol{\ell}}^{-1} \boldsymbol{\Delta} \tag{4.25}$$

and $\boldsymbol{l}_{\max}$ is its corresponding eigenvector. The index plot of $\boldsymbol{l}_{\max}$ may show which cases (observations) have a high influence on the estimate of $\boldsymbol{\theta} = (\alpha, \boldsymbol{\eta}^\top)^\top$.

If the interest is only on the vector $\boldsymbol{\eta}$, the normal curvature at the direction vector $\boldsymbol{l}$ is given by

$$C_l(\boldsymbol{\eta}) = 2\left|\boldsymbol{l}^\top \boldsymbol{\Delta}^\top \left(\ddot{\boldsymbol{\ell}}^{-1} - \boldsymbol{B}_1\right) \boldsymbol{\Delta}\boldsymbol{l}\right|; \tag{4.26}$$

see details in Cook (1986). The index plot of the largest eigenvector of $\boldsymbol{\Delta}^\top(\ddot{\boldsymbol{\ell}}^{-1} - \boldsymbol{B}_1)\boldsymbol{\Delta}$ may reveal the most influential observations on $\boldsymbol{\eta}$. Specifically, in the case of Birnbaum–Saunders regression models, we have

$$\boldsymbol{B}_1 = \begin{pmatrix} \ddot{\ell}_{\alpha\alpha}^{-1} & \boldsymbol{0} \\ \boldsymbol{0} & \boldsymbol{0} \end{pmatrix},$$

where $\ddot{\ell}_{\alpha\alpha}$ must be obtained from its corresponding expression according to the considered model. In addition, local influence of the observations on $\hat{\alpha}$ may be assessed by considering the index plot of $\boldsymbol{l}_{\max}$ of the matrix $\boldsymbol{\Delta}^\top(\ddot{\boldsymbol{\ell}}^{-1} - \boldsymbol{B}_2)\boldsymbol{\Delta}$, whose normal curvature at the direction vector $\boldsymbol{l}$ is given by

$$C_l(\alpha) = 2\left| \boldsymbol{l}^\top \boldsymbol{\Delta}^\top \left(\ddot{\boldsymbol{\ell}}^{-1} - \boldsymbol{B}_2 \right) \boldsymbol{\Delta} \boldsymbol{l} \right|, \tag{4.27}$$

where

$$\boldsymbol{B}_2 = \begin{pmatrix} \mathbf{0} & \mathbf{0} \\ \mathbf{0} & \ddot{\boldsymbol{\ell}}_{\eta\eta}^{-1} \end{pmatrix} = \begin{pmatrix} \mathbf{0} & \mathbf{0} \\ \mathbf{0} & (\boldsymbol{X}^\top \boldsymbol{V} \boldsymbol{X})^{-1} \end{pmatrix},$$

with $\boldsymbol{V}$ having elements as given in Equation (4.15).

As mentioned, besides the direction vector of maximum normal curvature $\boldsymbol{l}_{\max}$, it is possible to consider the vector $\boldsymbol{l} = \boldsymbol{e}_{iq}$, which corresponds to the direction vector of the case i, where $\boldsymbol{e}_{iq}$ is a $q \times 1$ vector of zeros with an one (1) at the ith position. Then, the total local influence of the case i is given by

$$C_i = 2|b_{ii}|, \quad i = 1, 2, \ldots, n, \tag{4.28}$$

where b_{ii} is the ith diagonal element of $\boldsymbol{B}$ given in Equation (4.25). Thus, the case i is considered as potentially influential if

$$C_i > 2\,\overline{C}, \quad \overline{C} = \frac{\sum_{i=1}^{n} C_i}{n}. \tag{4.29}$$

The index plot of C_i with benchmark given by $2\bar{C}$ is known as the total local influence method.

Below, we compute the elements of the $(p + 2) \times n$ perturbation matrix $\boldsymbol{\Delta}$ given by

$$\boldsymbol{\Delta} = (\Delta_{ij})_{(p+2)\times n} = \left(\frac{\partial^2 \ell(\boldsymbol{\theta}|\boldsymbol{\omega})}{\partial\theta_i \partial\omega_j} \right), \quad i = 1, \ldots, n, \ j = 1, \ldots, p + 2, \tag{4.30}$$

for three perturbation schemes according to the considered model. The matrix given in Equation (4.30) can be partitioned as

$$\boldsymbol{\Delta} = \begin{pmatrix} \boldsymbol{\Delta}_\alpha \\ \boldsymbol{\Delta}_\eta \end{pmatrix},$$

where $\boldsymbol{\Delta}_\alpha$ is an $n \times 1$ vector and $\boldsymbol{\Delta}_\eta$ is a $p \times n$ matrix. Every expression for $\boldsymbol{\Delta}$ given below must be evaluated at $\hat{\boldsymbol{\theta}}$. For these expressions, $h_Y(\cdot)$ is the failure rate of the log-Birnbaum–Saunders distribution given in Equation (4.5), ξ_{i1} and ξ_{i2} are given in Equation (4.17), $\boldsymbol{\omega} = (\omega_1, \ldots, \omega_n)^\top$ is the weight vector, $\boldsymbol{\omega}_0 = (1, \ldots, 1)^\top$ is the nonperturbation vector, $\mu_i = \mathrm{E}(Y_i)$ and Y_i follows a log-Birnbaum–Saunders distribution, for all $i = 1, \ldots, n$.

Local influence for the Birnbaum–Saunders regression model in the case of uncensored data is detailed below.

Case-weight perturbation

Under this scheme, the relevant part of the corresponding perturbed log-likelihood function is given by

$$\ell(\boldsymbol{\theta}|\boldsymbol{\omega}) = \sum_{i=1}^{n} \omega_i \log(\xi_{i1}) - \frac{1}{2}\sum_{i=1}^{n} \omega_i \xi_{i2}^2, \tag{4.31}$$

for $0 \leq \omega_i \leq 1$. In this case, the perturbation matrix $\boldsymbol{\Delta}$ has elements

$$\boldsymbol{\Delta}_\alpha = (\phi_1, \ldots, \phi_n), \quad \boldsymbol{\Delta}_\eta = \boldsymbol{X}^\top \text{diag}\{b_1, \ldots, b_n\},$$

where

$$\phi_i = -\frac{1}{\alpha}\left(1 + \xi_{i2}^2\right), \quad b_i = \frac{1}{2}\left(\xi_{i1}\xi_{i2} - \frac{\xi_{i2}}{\xi_{i1}}\right), \; i = 1, 2, \ldots, n,$$

and ξ_{i1}, ξ_{i2} given in Equation (4.10).

Response perturbation

Consider the regression model given in Equation (4.8) and assume that each observation y_i is perturbed as

$$y_{i\omega} = y_i + \omega_i S_y, \quad i = 1, \ldots, n, \tag{4.32}$$

where S_y is a scale factor that can correspond to the sample standard deviation of Y. In this case, the relevant part of the corresponding perturbed log-likelihood function is given by

$$\ell(\boldsymbol{\theta}|\boldsymbol{\omega}) = \sum_{i=1}^{n} \log(\xi_{i1\omega 1}) - \frac{1}{2}\sum_{i=1}^{n} \xi_{i2\omega 1}^2, \tag{4.33}$$

where $\xi_{i1\omega 1}$ and $\xi_{i2\omega 1}$ are given implicitly in Equation (4.10), with y_i being replaced by $y_{i\omega}$. Then, the corresponding perturbation matrix $\boldsymbol{\Delta}$ has elements

$$\boldsymbol{\Delta}_\alpha = (\varrho_1, \ldots, \varrho_n), \quad \boldsymbol{\Delta}_\eta = \boldsymbol{X}^\top \text{diag}\{d_1, \ldots, d_n\}, \tag{4.34}$$

where

$$\varrho_i = \frac{S_y\, \xi_{i1}\, \xi_{i2}}{\alpha}, \quad i = 1, 2, \ldots, n,$$

and

$$d_i = S_y\left(\frac{1}{\alpha^2}\cosh(y_i - \mu_i) - \frac{1}{4}\left(\text{sech}\left(\frac{y_i - \mu_i}{2}\right)\right)^2\right). \tag{4.35}$$

Explanatory variable perturbation

Consider now an additive perturbation on a continuous explanatory variable, namely X_p, such that its observed value is

$$x_{ip\omega} = x_{ip} + \omega_i S_x, \quad i = 1, 2, \ldots, n,$$

where S_x is a scale factor that can correspond to the sample standard deviation of X_p. Then, the relevant part of the corresponding perturbed log-likelihood function is given by

$$\ell(\boldsymbol{\theta}|\boldsymbol{\omega}) = \sum_{i=1}^{n} \log(\xi_{i1\omega 2}) - \frac{1}{2}\sum_{i=1}^{n} \xi_{i2\omega 2}^2, \tag{4.36}$$

where $\xi_{i1\omega 2}$ and $\xi_{i2\omega 2}$ are given implicitly in Equation (4.10), replacing in μ_i, x_{ip} with $x_{ip\omega}$. In this case, the corresponding perturbation matrix $\boldsymbol{\Delta}$ has elements

$$\boldsymbol{\Delta}_\alpha = (\varphi_1, \ldots, \varphi_n), \quad \boldsymbol{\Delta}_\eta = \frac{S_x \eta_p}{2}\left(\boldsymbol{X}^\top \mathrm{diag}\{k_1, \ldots, k_n\} + \begin{pmatrix} \boldsymbol{0} \\ \boldsymbol{q}^\top \end{pmatrix}\right),$$

where $\boldsymbol{q} = (q_1, \ldots, q_n)^\top$ and

$$\varphi_i = -\frac{S_x\, \eta_p\, \xi_{i1}\, \xi_{i2}}{\alpha}, \quad i = 1, 2, \ldots, n,$$

$$k_i = -\frac{4}{\alpha^2}\cosh(y_i - \mu_i) + \frac{1}{2}\left(\mathrm{sech}\left(\frac{y_i - \mu_i}{2}\right)\right)^2, \tag{4.37}$$

$$q_i = \frac{4}{\alpha^2}\sinh(y_i - \mu_i) - \tanh\left(\frac{y_i - \mu_i}{2}\right). \tag{4.38}$$

Local influence for the Birnbaum–Saunders regression model in the case of censored data is detailed below.

Case-weight perturbation

Under this scheme, the relevant part of the corresponding perturbed log-likelihood function is given by

$$\ell(\boldsymbol{\theta}\,|\,\boldsymbol{\omega}) = \sum_{i\in D} \omega_i \left(\log(\xi_{i1}) - \frac{\xi_{i2}^2}{2}\right) + \sum_{i\in C} \omega_i \log(\Phi(-\xi_{i2})), \tag{4.39}$$

for $0 \le \omega_i \le 1$. In this case, the perturbation matrix $\boldsymbol{\Delta}$ has elements

$$\boldsymbol{\Delta}_\alpha = (a_1, \ldots, a_n), \quad \boldsymbol{\Delta}_\eta = \boldsymbol{X}^\top \mathrm{diag}\{b_1, \ldots, b_n\},$$

where

$$a_i = \begin{cases} -\frac{1}{\alpha} + \frac{\xi_{i2}^2}{\alpha}, & i \in D; \\ \frac{\xi_{i2}}{\alpha} h_Y(\xi_{i2}; \boldsymbol{\theta}), & i \in C; \end{cases} \quad b_i = \begin{cases} \frac{1}{2}\left(\xi_{i1}\xi_{i2} - \frac{\xi_{i2}}{\xi_{i1}}\right), & i \in D; \\ \frac{\xi_{i1}}{2} h_Y(\xi_{i2}; \boldsymbol{\theta}), & i \in C. \end{cases} \tag{4.40}$$

Response perturbation

Consider the regression model given in Equation (4.16) and assume that each y_i is perturbed as

$$y_{i\omega} = y_i + \omega_i S_y, \quad i = 1, 2, \ldots, n,$$

where S_y can correspond to the sample standard deviation of Y. In this case, the relevant part of the corresponding perturbed log-likelihood function is given by

$$\ell(\boldsymbol{\theta} \,|\, \boldsymbol{\omega}) = \sum_{i \in D} \left(\log(\xi_{i1\omega_1}) - \frac{\xi_{i2\omega_1}^2}{2} \right) + \sum_{i \in C} \log\left(\Phi(-\xi_{i2\omega_1})\right), \tag{4.41}$$

where $\xi_{i1\omega 1}$ and $\xi_{i2\omega 1}$ are given implicitly in Equation (4.10), with y_i being replaced by $y_{i\omega}$. Then, the corresponding perturbation matrix $\boldsymbol{\Delta}$ has elements

$$\boldsymbol{\Delta}_\alpha = (c_1, \ldots, c_n), \quad \boldsymbol{\Delta}_\eta = \boldsymbol{X}^\top \mathrm{diag}\{d_1, \ldots, d_n\},$$

where

$$c_i = \begin{cases} \frac{S_y}{\alpha} \xi_{i1}\xi_{i2}, & i \in D; \\ \frac{S_y}{2\alpha}\left(\xi_{i1} h_Y(\xi_{i2}; \boldsymbol{\theta}) + \xi_{i1}\xi_{i2} h'(\xi_{i2})\right), & i \in C; \end{cases}$$

$$d_i = \begin{cases} S_y\left(\frac{1}{\alpha^2}\cosh(y_i - \mu_i) - \frac{1}{4}\left(\mathrm{sech}\left(\frac{y_i - \mu_i}{2}\right)\right)^2\right), & i \in D; \\ \frac{S_y}{4}\left(\xi_{i2} h_Y(\xi_{i2}; \boldsymbol{\theta}) + \xi_{i1}^2 h'_Y(\xi_{i2}; \boldsymbol{\theta})\right), & i \in C. \end{cases}$$

Explanatory variable perturbation

Consider once again an additive perturbation on a particular continuous explanatory variable X_p such that its observed value is

$$x_{it\omega} = x_{it} + \omega_i S_x, \quad i = 1, 2, \ldots, n,$$

where S_x can correspond to the sample standard deviation of X_p. In this case, the relevant part of the corresponding perturbed log-likelihood function is given by

$$\ell(\boldsymbol{\theta} \,|\, \boldsymbol{\omega}) = \sum_{i \in D} \left(\log(\xi_{i1\omega_2}) - \frac{1}{2}\xi_{i2\omega_2}^2 \right) + \sum_{i \in C} \log\left(\Phi(-\xi_{i2\omega_2})\right), \tag{4.42}$$

where $\xi_{i1\omega2}$ and $\xi_{i2\omega2}$ are given implicitly in Equation (4.10), replacing in μ_i, x_{ip} with $x_{ip\omega}$. Then, the elements of the corresponding perturbation matrix $\boldsymbol{\Delta}$ assume the form $\boldsymbol{\Delta}_\alpha = (\phi_1, \ldots, \phi_n)$, where

$$\phi_i = \begin{cases} -\frac{2}{\alpha^3} S_x \eta_t \sinh(y_i - \mu_i), & i \in D; \\ -\frac{\eta_t S_x}{2\alpha} \left(\xi_{i1}\, h_Y(\xi_{i2}; \boldsymbol{\theta}) + \frac{2}{\alpha^2} \sinh(y_i - \mu_i)\, h'_Y(\xi_{i2}; \boldsymbol{\theta}) \right), & i \in C; \end{cases}$$

and $\boldsymbol{\Delta}_\eta = (\Delta_{\eta_{ij}})$, where, for $j \neq t$,

$$\Delta_{\eta_{ij}} = \begin{cases} S_x \eta_t x_{ij} \left(\frac{1}{4} \left(\operatorname{sech}\left(\frac{y_i - \mu_i}{2} \right) \right)^2 - \frac{1}{\alpha^2} \cosh(y_i - \mu_i) \right), & i \in D; \\ -\frac{S_x \eta_t x_{ij}}{4} \left(\xi_{i2}\, h_Y(\xi_{i2}; \boldsymbol{\theta}) + \xi_{i1}^2\, h'_Y(\xi_{i2}; \boldsymbol{\theta}) \right), & i \in C; \end{cases}$$

whereas if $j = t$, one has

$$\Delta_{\eta_{it}} = \begin{cases} S_x \eta_t x_{it} \left(\frac{1}{4} \left(\operatorname{sech}\left(\frac{y_i - \mu_i}{2} \right) \right)^2 - \frac{1}{\alpha^2} \cosh(y_i - \mu_i) \right) & \\ + S_x \left(\frac{1}{\alpha^2} \sinh(y_i - \mu_i) - \frac{1}{2} \tanh\left(\frac{y_i - \mu_i}{2} \right) \right), & i \in D; \\ -\frac{S_x \eta_t x_{it}}{4} \left(\xi_{i2}\, h_Y(\xi_{i2}; \boldsymbol{\theta}) + \xi_{i1}^2\, h'_Y(\xi_{i2}; \boldsymbol{\theta}) \right) & \\ + \frac{S_x}{2} \xi_{i1}\, h_Y(\xi_{i2}; \boldsymbol{\theta}), & i \in C. \end{cases}$$

Noncensored case perturbation

Suppose that only the noncensored cases are perturbed, which means to assume $\omega_i = 1$, for $i \in C$, in expression given in Equation (4.39). Thus, the quantities a_i and b_i given in Equation (4.40) reduce, respectively, to

$$a_i = \begin{cases} -\frac{1}{\alpha} + \frac{\xi_{i2}^2}{\alpha}, & i \in D; \\ 0, & i \in C; \end{cases} \qquad b_i = \begin{cases} \frac{1}{2} \left(\xi_{i1}\, \xi_{i2} - \frac{\xi_{i2}}{\xi_{i1}} \right), & i \in D; \\ 0, & i \in C. \end{cases}$$

Censored case perturbation

Suppose now that only the censored cases are perturbed, that is, $\omega_i = 1$ for $i \in D$, in expression given in Equation (4.39). In this case, the quantities a_i and b_i given in Equation (4.40) reduce, respectively, to

$$a_i = \begin{cases} 0, & i \in D; \\ \frac{\xi_{i2}}{\alpha}\, h_Y(\xi_{i2}; \boldsymbol{\theta}), & i \in C; \end{cases} \qquad b_i = \begin{cases} 0, & i \in D; \\ \frac{\xi_{i1}}{2}\, h_Y(\xi_{i2}; \boldsymbol{\theta}), & i \in C. \end{cases} \tag{4.43}$$

Generalized leverage

Let $\ell(\boldsymbol{\theta})$ denote the corresponding log-likelihood function. In addition, let $\hat{\boldsymbol{\theta}}$ be the maximum likelihood estimate of $\boldsymbol{\theta}$ and $\boldsymbol{\mu}$ the expectation of $\boldsymbol{Y} = (Y_1, Y_2, \ldots, Y_n)^\top$. Then, $\hat{\boldsymbol{y}} = \boldsymbol{\mu}(\hat{\boldsymbol{\theta}})$ is the predicted response vector.

The concept of leverage consists of evaluating the influence of the observed response variable, say y_i, for $i = 1, 2, \ldots, n$, on its predicted value, say $\hat{y}_i$; see, for example, Cook and Weisberg (1982) and Wei et al. (1998). This influence may be well represented by the derivative $\partial\hat{y}_i/\partial y_i$ that is equal to p_{ii} in the normal linear model, where p_{ii} is the ith principal diagonal element of the projection matrix

$$\boldsymbol{P} = \boldsymbol{X}\left(\boldsymbol{X}^\top\boldsymbol{X}\right)^{-1}\boldsymbol{X}^\top$$

and $\boldsymbol{X}$ is the design matrix of the model. Extensions of the leverage concept to more general regression models have been provided, for example, by St. Laurent and Cook (1992) and Wei et al. (1998), when $\boldsymbol{\theta}$ is unrestricted, and by Paula (1999) when restricted.

From Wei et al. (1998), we have that the $n \times n$ matrix $(\partial\hat{\boldsymbol{y}}/\partial\boldsymbol{y}^\top)$ of generalized leverage may be expressed by

$$\mathrm{GL}(\boldsymbol{\theta}) = \boldsymbol{D}_\theta\left(-\ddot{\boldsymbol{\ell}}\right)^{-1}\ddot{\boldsymbol{\ell}}_{\theta y}, \tag{4.44}$$

evaluated at $\hat{\boldsymbol{\theta}}$, where

$$\boldsymbol{D}_\theta = \frac{\partial\boldsymbol{\mu}}{\partial\boldsymbol{\theta}^\top} = (\boldsymbol{X}, \boldsymbol{0}), \quad \ddot{\boldsymbol{\ell}}_{\theta y} = \frac{\partial^2\ell(\boldsymbol{\theta})}{\partial\boldsymbol{\theta}\partial\boldsymbol{y}^\top} = \begin{pmatrix} \frac{\partial^2\ell(\boldsymbol{\theta})}{\partial\boldsymbol{\eta}\partial\boldsymbol{y}^\top} \\ \frac{\partial^2\ell(\boldsymbol{\theta})}{\partial\alpha\partial\boldsymbol{y}^\top} \end{pmatrix} = \begin{pmatrix} \ddot{\boldsymbol{\ell}}_{\eta y} \\ \ddot{\boldsymbol{\ell}}_{\alpha y} \end{pmatrix},$$

with $\ddot{\boldsymbol{\ell}}_{\theta y}$ being the corresponding Hessian matrix. An alternative expression for the generalized leverage is given by

$$\mathrm{GL}(\boldsymbol{\theta}) = S_y\boldsymbol{\Psi}\boldsymbol{V} + rS_y\boldsymbol{\Psi}\boldsymbol{k}\boldsymbol{k}^\top(\boldsymbol{I}_n - \boldsymbol{\Psi}\boldsymbol{V}), \tag{4.45}$$

where

$$\boldsymbol{\Psi} = \boldsymbol{X}(\boldsymbol{X}^\top\boldsymbol{V}\boldsymbol{X})^{-1}\boldsymbol{X}^\top,$$

$\boldsymbol{k}$ and $\boldsymbol{V}$ are given in Equations (4.15), S_y is a scale factor associated with the perturbation scheme of the response variable and

$$r = \frac{1}{\boldsymbol{k}^\top\boldsymbol{\Psi}\boldsymbol{k} + 2n/\alpha^2}. \tag{4.46}$$

Index plots of GL_{ii} versus i may reveal the case i has a high influence on its predicted value.

Residual analysis

To study departures from the error assumptions as well as presence of atypical observations, we consider two kinds of residuals: deviance component and martingale-type. The interested reader is referred to, for example,

McCullagh and Nelder (1989) for the deviance component residual, Barlow and Prentice (1988) and Therneau et al. (1990) for the martingale-type residual, and Ortega et al. (2003) for both of them.

In general, the deviance component residual is defined as

$$r_{\mathrm{DC}_i} = \mathrm{sign}(y_i - \hat{\mu}_i)\sqrt{2\,(\ell_i(\hat{\boldsymbol{\theta}}_{\mathrm{s}}) - \ell_i(\hat{\boldsymbol{\theta}}))}, \quad i = 1, 2, \ldots, n, \tag{4.47}$$

where $\ell_i(\cdot)$ is the log-likelihood for the case i, $\hat{\boldsymbol{\theta}}_{\mathrm{s}}$ is the maximum likelihood estimate of $\boldsymbol{\theta}$ under the saturated model (with n parameters), $\hat{\boldsymbol{\theta}}$ is the maximum likelihood estimate of $\boldsymbol{\theta}$ under the model of interest (with $p + 2$ parameters), $\hat{\mu}_i = \mathrm{E}(Y_i)$, and sign($z$) denotes the sign of z. Davison and Gigli (1989) defined from Equation (4.47) the deviance component residual for censored data as

$$r_{\mathrm{DC}_i} = \mathrm{sign}(y_i - \hat{\mu}_i)\sqrt{-2\log\left(\hat{R}_Y(y_i; \boldsymbol{\theta})\right)}, \quad i = 1, 2, \ldots, n, \tag{4.48}$$

where $\hat{R}_Y(\cdot)$ is the maximum likelihood estimate of the corresponding reliability function; see, for example, Ortega (2001).

Therneau et al. (1990) introduced the deviance component residual in counting processes basically using martingale residuals, which are skewed and have a maximum value at $+1$ and a minimum value at $-\infty$. For censored data, the martingale residual can be expressed by

$$r_{\mathrm{M}_i} = \delta_i + \log\left(\hat{R}_Y(y_i; \boldsymbol{\theta})\right), \quad i = 1, 2, \ldots, n, \tag{4.49}$$

where $\delta_i = 0$ indicates if the case i is censored, $\delta_i = 1$ if the case i is uncensored and once again $\hat{R}(\cdot)$ is the maximum likelihood estimate with censored data of the corresponding reliability function; see, for example, Klein and Moeschberger (1997) and Ortega et al. (2003).

The deviance component residual proposed by Therneau et al. (1990) is a transformation of the martingale residual to attenuate the skewness. This transformation was motivated by the deviance component residual found in generalized linear models. In particular, the deviance component residual for the Cox proportional hazard model with no time-dependent explanatory variables is named the martingale-type residual and described as

$$r_{\mathrm{MT}_i} = \mathrm{sign}(r_{\mathrm{M}_i})\sqrt{-2\left(r_{\mathrm{M}_i} + \delta_i \log(\delta_i - r_{\mathrm{M}_i})\right)}, \quad i = 1, 2, \ldots, n, \tag{4.50}$$

where r_{M_i} is the martingale residual defined in Equation (4.49). Although the martingale-type residual expressed in Equation (4.50) is not a deviance component of log-Birnbaum–Saunders regression models, we use it in the sequel as a transformation of the martingale residual to have residuals symmetrically distributed around zero.

Ortega (2001) suggested to standardize the deviance component and martingale-type residuals for censored data as

$$r^*_{\mathrm{DC}i} = \frac{r_{\mathrm{DC}i}}{\sqrt{1 - \mathrm{GL}_{ii}}}, \quad r^*_{\mathrm{MT}_i} = \frac{r_{\mathrm{MT}_i}}{\sqrt{1 - \mathrm{GL}_{ii}}}, \qquad i = 1, 2, \ldots, n,$$

respectively, with GL_{ii} being the ith principal diagonal element of the generalized leverage matrix given in Equation (4.44) evaluated at $\hat{\boldsymbol{\theta}}$.

Considering $\alpha < 2$ fixed or known, we have that the deviance component residual for Birnbaum–Saunders regression models is given by

$$r_{\mathrm{DC}_i} = \begin{cases} \mathrm{sign}(y_i - \hat{\mu}_i)\sqrt{-2\log\left(\cosh\left(\frac{y_i-\hat{\mu}_i}{2}\right)\right) + \frac{2}{\hat{\alpha}^2}\left(\sinh\left(\frac{y_i-\hat{\mu}_i}{2}\right)\right)^2}, \\ \quad i \in D; \\ \mathrm{sign}(y_i - \hat{\mu}_i)\sqrt{-2\log\left(\Phi\left(-\frac{2}{\hat{\alpha}}\sinh\left(\frac{y_i-\hat{\mu}_i}{2}\right)\right)\right)}, \\ \quad i \in C. \end{cases}$$

In addition, the martingale residual for Birnbaum–Saunders regression models assumes the form

$$r_{\mathrm{M}_i} = \delta_i + \log\left(\Phi\left(-\frac{2}{\hat{\alpha}}\sinh\left(\frac{y_i - \hat{\mu}_i}{2}\right)\right)\right), \qquad i = 1, 2, \ldots, n.$$

CHAPTER 5

Goodness of Fit for the Birnbaum–Saunders Distribution

Abstract

In this chapter, goodness-of-fit methods and criteria of model selection for the Birnbaum–Saunders distribution are presented. First, we provide general aspects on good of fit, detail theoretical arguments often used for considering life distributions as candidates to model lifetime data, and introduce the total time on test. Second, we provide good-of-fit tools based on moments. Third, we discuss model selection procedures based on loss of information, such as Akaike, Schwarz's Bayesian, and Hannan–Quinn criteria and the Bayes factor, which are useful to statistically compare a model to other models. This chapter is finished describing probability plots and good-of-fit methods, mainly based on Kolmogorov-Smornov and Michael tests, which allow us to contruct acceptance bands inside the corresponding probability plots.

Keywords: Akaike, Schwarz and Hannan–Quinn information criteria, beta function, censored data, cumulative distribution function, empirical distribution function, failure rate, goodness-of-fit methods, Kolmogorov-Smornov test, maximum likelihood method, moments, order statistics, probability integral transform, probability–probability, quantile–quantile and stabilized-probability plots, p-value, standard normal distribution, total time on test, uniform distribution.

5.1 INTRODUCTION

One of the most complex problems in parametric statistical methods is the identification of the distribution that can best fit the data. Some general methods for goodness of fit are the probability–probability (PP) and quantile–quantile (QQ) plots and the Kolmogorov–Smirnov (KS) test. Also, some model selection criteria can be used to select an appropriate statistical distribution. Next, we explain how a distribution must be selected. Then, in addition to the graphical method introduced in Section 3.4 of Chapter 3, in order to provide goodness-of-fit tools for the Birnbaum–

The Birnbaum–Saunders Distribution. http://dx.doi.org/10.1016/B978-0-12-803769-0.00005-4

Saunders distribution, we present methods based on moments and probability plots. The interested reader is referred to the classic book by D'Agostino and Stephens (1986) for different goodness-of-fit methods.

A very important aspect in parametric lifetime analysis is the identification of the appropriate life distribution. Most of the probabilistic models used to describe lifetime data are chosen by using one or more of the following aspects:

(F1) A theoretical argument for the mechanism of failure of the item.
(F2) A model that has previously been used with a successful result.
(F3) A model that fits to the data well.

For more details about (F1) to (F3), the interested reader is referred to Tobias (2004, Section 8.2.1).

Based exclusively on the statistical view of point (F3), a frequent problem is that the life tests generate little data that represent the failure of the item. This is generally due to a time horizon of the life test that is relatively short or an insufficient amount of money to wait until all the items fail. In lifetime data, it has been verified that some life distribution families of two parameters, that have certain flexibility, fit reasonably well in the central region, as there is usually a reasonable amount of data in the central part of the distribution. However, in lifetime data analysis sometimes the interest is in the low or high quantiles of the distribution, for example, to establish lethal doses, warranties, safe-life reserves, or policies of maintenance. Nevertheless, the fit of the life distribution at the tails is often not good, because few data are available at the extremes of their empirical distribution. This conducts to a wide discrepancy among some life distributions of similar characteristics. Such a situation hinders the identification of an appropriate life distribution through visual and goodness-of-fit methods. Then, under these conditions of few data at the extremes, it is problematic to consider these methods to decide what distribution must be used. Thus, it is better to propose a life distribution using a theoretical argument; see Section 1.4 in Chapter 1. Some examples of these arguments are the following:

(i) Consider a hypothetical system that is formed by one or more identical components. One of these components is designated to be the primary unit and the others as secondary units. The system starts operating with the primary unit until its failure, time at which the first of the secondary units begins to operate. When the first of the secondary units does not work, the next unit enters in operation, and so on until all the secondary units have failed. Therefore, the system operates during as long as support units exist. Thus, the lifetime of the system is the sum of the lifetimes of each one of the units that conform it. Suppose that the units do not have wear with the use and that the time of failure of the units are independent. Under this scenario, the lifetime of each unit must be modeled by the exponential distribution and therefore the lifetime of the system is governed by the gamma distribution. This example is analogous to an argument of "wear-out of type phase" with exponential lifetimes for each phase, which also justifies a gamma or Erlang distribution.

(ii) It is well-known that an argument of type "extreme values" justifies the use of the Weibull distribution.

(iii) A "multiplicative degradation" argument justifies the log-normal distribution.

(iv) An argument of "first step time" justifies the use of the inverse Gaussian distribution.

(v) An argument of "fatigue" or "cumulative damage" justifies the use of the Birnbaum–Saunders distribution.

An analysis of the probability density function is not always appropriate. Another way to state the suitability of a life distribution can be based on lifetime indicators; see, for example, Cox and Oakes (1984, pp. 24–26). Then, one can consider the following:

(G1) To study the failure rate of T, $h_T(t)$, or its logarithm, $\log(h_T(t))$, against t or $\log(t)$.

(G2) To analyze the cumulative failure rate, $H_T(t)$, or its logarithm, against t or $\log(t)$.

In practice, when lifetime data are analyzed and we want to propose a distribution for modeling them, one often construct a histogram. As it is well know, the histogram is an empirical approximation of the probability density function of the data, which is simple and easy to contruct. However, it is always convenient to take a look at the failure rate of the data, because distributions with similar probability density functions could have different failure rates. This occurs, for example, among the gamma, Weibull distributions and Birnbaum–Saunders, inverse Gaussian, log-normal distributions. The problem here is that approximating empirically the failure rate is not an easy task. Thus, a tool that is used to facilitate this task corresponds to the total time on test (TTT) plot. This plot is related to (G1) and (G2) and allows us to detect the shape of the failure rate of a random variable and, as consequence, of the distribution that the data could follow. The TTT function of the random variable T is given by

$$H_T^{-1}(u) = \int_0^{F_T^{-1}(u)} (1 - F_T(y))\,\mathrm{d}y,$$

whereas its scaled version is defined as

$$W_T(u) = \frac{H_T^{-1}(u)}{H_T^{-1}(1)}, \quad 0 \leq u \leq 1,$$

where once again $F_T^{-1}(\cdot)$ is the inverse function of the cumulative distribution function of T. A graph of the theoretical scaled TTT curve is depicted in Figure 5.1. Now, $W_T(\cdot)$ can be approximate allowing us to construct the empirical scaled TTT curve plotting the points $(k/n, W_n(k/n))$, where

$$W_n\left(\frac{k}{n}\right) = \frac{\sum_{i=1}^{k} t_{i:n} + (n-k)t_{k:n}}{\sum_{i=1}^{n} t_{i:n}}, \quad k = 1, \ldots, n,$$

with $t_{i:n}$ being the observed value of the ith order statistic, for $i = 1, \ldots, n$. Thus, a TTT function that is concave (or convex) corresponds to the

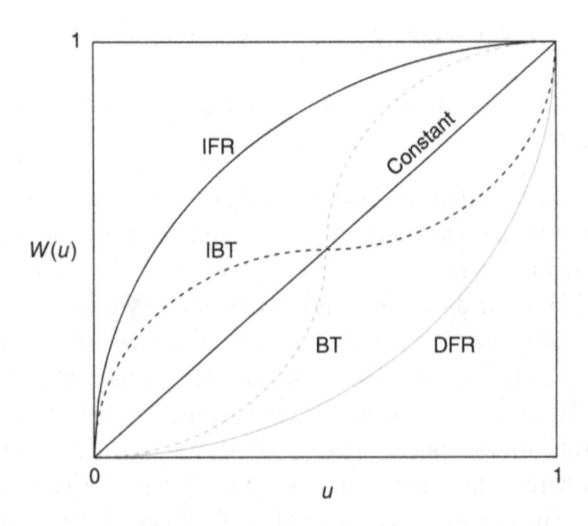

Figure 5.1 Theoretical TTT plots.

increasing failure rate (IFR) (or decreasing failure rate, DRF) class. A TTT function that is concave (convex) at the left from the change point and convex (concave) at the right from the change point corresponds to a bathtub (BT) (or inverse bathtub, IBT) failure rate. A TTT plot expressed by a straight line is an indication that the underlying distribution is exponential, which, as it is well known, has a constant failure rate. The interested reader is referred to Aarset (1987) for more details about TTT plot.

In spite of all previously mentioned in this section, we cannot only consider the use of these theoretical arguments and lifetime indicators. In summary, whatever the justification used for choosing one or several probability models as candidates, the life distribution to be employed for describing the data must be logical and supported by visual tools and statistical tests of goodness of fit to the data and/or criteria for model selection.

5.2 GOODNESS OF FIT BASED ON MOMENTS

The use of goodness-of-fit graphical methods based on moments are particularly important when confronting data obtained from several sampling points, for example, air pollution data. These methods are especially useful in distributions within the scale family, such as the Birnbaum–Saunders, gamma, and log-normal distributions, since they produce dimensionless ratios functionally independent of the scale parameter.

Goodness-of-fit methods based on moments consist of developing rectangular charts with coordinate axes computed from moments. For example, the β_1–β_2 chart has coordinate axes (β_1, β_2), that is, based on the coefficients of skewness and kurtosis, where $\beta_1 = (\alpha_3(T))^2$ and $\beta_2 = \alpha_2(T)$ according to Equations (2.24) and (2.25), respectively, given by

$$\beta_2 = \frac{\mu_2\,\mu_4}{(\mu_3)^2}\,\beta_1,$$

with μ_2, μ_3, and μ_4 being the second, third, and fourth central moments, respectively. However, Tsukatani and Shighemitsu (1980) indicated that the β_1–β_2 chart is not always appropriate because the fourth moment lacks sufficient accuracy when it is used in some types of data, for example, in air pollutant data. Thus, they proposed to replace the β_1–β_2 chart with an alternative chart characterized by a curve on the plane δ_2–δ_3, that is, based on the coefficients of variation and skewness, where $\delta_2 = \mu_2/\mu^2$ and $\delta_3 = \mu_3/(\mu\,(\mu_2)^2)$, such that

$$\delta_3 = \frac{\mu\,\mu_3}{(\mu_2)^2}\,\delta_2, \tag{5.1}$$

with μ being the expected value of the distribution. A large value of δ_3 given in Equation (5.1) indicates a tail of the distribution that is relatively heavy. Nevertheless, in general, the third moment, μ_3, has a large sampling error for heavy-tailed distributions. Thus, another alternative could be to consider charts based on fractional moments, for example, of order $1/2$ or $3/2$. This means that, in such a case, we can use the chart based on

$$\gamma = \frac{\sigma}{\mu} \quad \text{versus} \quad \gamma_{1/2} = \frac{\mu_{1/2}}{\sigma^{1/2}} \quad \text{or} \quad \gamma_{3/2} = \frac{\mu_{3/2}}{\sigma^{3/2}},$$

where $\sigma = \sqrt{\mu_2}$. In order to highlight the left tail, we can use $\gamma_{-1/2} = \mu_{-1/2}/\sigma^{1/2}$ instead of $\gamma_{1/2} = \mu_{1/2}/\sigma^{1/2}$. The δ_2–δ_3 chart defined in Equation (5.1) is analogous to the γ–γ_3 chart proposed in Cox and Oakes (1984, pp. 26–27), where $\gamma = \gamma(T) = \sqrt{\delta_2}$ and $\gamma_3 = \sqrt{\alpha_3(T)}$ are given in Equations (2.22) and (2.24), respectively. The interested reader is referred to the classic references Elderton (1927) and Ord (1972) for more details about goodness-of-fit charts based on moments.

Johnson et al. (1995, p. 655) mentioned the analysis of rainfall characteristics of Hiroshima city based on the δ_2–δ_3 diagram and that the distribution of periods of continuous rainfall is best fitted by a Birnbaum–Saunders distribution; see also Seto et al. (1993, 1995). Meeker and Escobar (1998, p. 110) commented about the fitting lines based on moments for the Birnbaum–Saunders distribution; see also Leiva et al. (2008d). Thus, based on Equations (2.22), (2.24), and (2.25), we can specify the β_1–β_2 and γ–γ_3 (or similarly δ_2–δ_3) charts for the BS distributions as

$$\beta_2 = \frac{3(5\alpha^2+4)(211\alpha^4+120\alpha^2+16)}{16\alpha^2(11\alpha^2+6)^2}\,\beta_1$$

and

$$\gamma_3 = \frac{4(11\alpha^2+6)(\alpha^2+2)}{(5\alpha^2+4)^2}\,\gamma.$$

5.3 MODEL SELECTION

In order to statistically compare a distribution to other distributions, we can use model selection procedures based on loss of information, such as Akaike (AIC), Schwarz's Bayesian (BIC), and Hannan–Quinn (HQIC) information criteria. These criteria allow us to compare models for the same data set and are given by

$$\begin{aligned} \text{AIC} &= -2\ell(\hat{\boldsymbol{\theta}}) + 2d, \\ \text{BIC} &= -2\ell(\hat{\boldsymbol{\theta}}) + d\log(n), \\ \text{HQIC} &= -2\ell(\hat{\boldsymbol{\theta}}) + 2d\log(\log(n)), \end{aligned}$$

where $\ell(\hat{\boldsymbol{\theta}})$ is the corresponding log-likelihood function for the parameter $\boldsymbol{\theta}$ associated with the model evaluated at $\boldsymbol{\theta} = \hat{\boldsymbol{\theta}}$, n is the sample size and d is the dimension of the parameter space.

AIC, BIC, and HQIC are based on a penalization of the likelihood function as the model becomes more complex, that is, as the model has more parameters. Thus, a model whose information criterion has a smaller value is better than another model whose criterion has a larger value. Because models with more parameters always provide a better fit, AIB, BIC, and HQIC allow us to compare models with different numbers of parameters due to the penalization incorporated in such criteria. This methodology is very general and can be applied even to non-nested models, that is, those models that are not particular cases of a more general model; see Kass and Raftery (1995), Raftery (1995), and Vilca et al. (2011).

Differences between two values of the information criteria are often not very noticeable. In these cases, the Bayes factor can be used to highlight such differences, if they exist. To define the Bayes factor, assume the data D belong to one of two hypothetical models, namely M_1 and M_2, according to probabilities $\mathrm{P}(D|\mathrm{M}_1)$ and $\mathrm{P}(D|\mathrm{M}_2)$, respectively. Given probabilities $\mathrm{P}(\mathrm{M}_1)$ and $\mathrm{P}(\mathrm{M}_2) = 1 - \mathrm{P}(\mathrm{M}_1)$, the data produce conditional probabilities $\mathrm{P}(\mathrm{M}_1|D)$ and $\mathrm{P}(\mathrm{M}_2|D) = 1 - \mathrm{P}(\mathrm{M}_1|D)$, respectively. Then, the Bayes factor that allows us to compare M_1 (model considered as correct) to M_2 (model to be contrasted to M_1) is given by

$$B_{12} = \frac{\mathrm{P}(D|\mathrm{M}_1)}{\mathrm{P}(D|\mathrm{M}_2)}. \tag{5.2}$$

Based on Equation (5.2), we can use the approximation

$$2\log(B_{12}) \approx 2(\ell(\hat{\boldsymbol{\theta}}_1) - \ell(\hat{\boldsymbol{\theta}}_2)) - (d_1 - d_2)\log(n), \tag{5.3}$$

where $\ell(\hat{\boldsymbol{\theta}}_k)$ is the log-likelihood function for the parameter $\boldsymbol{\theta}_k$ under the model M_k evaluated at $\boldsymbol{\theta}_k = \hat{\boldsymbol{\theta}}_k$, d_k is the dimension of $\boldsymbol{\theta}_k$, for $k = 1, 2$, and n is the sample size. Note that the approximation given in Equation (5.3) is computed subtracting the BIC value from the model M_2, given by $\mathrm{BIC}_2 = -2\ell(\boldsymbol{\theta}_2) + d_2 \log(n)$, to the BIC value of the model M_1, given by $\mathrm{BIC}_1 = -2\ell(\boldsymbol{\theta}_1) + d_1 \log(n)$. In addition, note that if model M_2 is a particular case of M_1, then the procedure corresponds to applying the likelihood ratio test. In such a case,

$$2\log(B_{12}) \approx \chi^2_{12} - \mathrm{df}_{12}\,\log(n),$$

where χ^2_{12} is the likelihood ratio statistic to test M_1 versus M_2 and $\mathrm{df}_{12} = d_1 - d_2$ are the degrees of freedom associated with the likelihood ratio test. Therefore, one can obtain the corresponding p-value from $2\log(B_{12}) \stackrel{\cdot}{\sim} \chi^2(d_1 - d_2)$, with $d_1 > d_2$. In general, the Bayes factor given in Equation (5.2) is informative because it presents ranges of values in which the degree of superiority of a model in comparison to other can be quantified. An interpretation based on the Bayes factor is displayed in Table 5.1; see Kass and Raftery (1995), Raftery (1995), and Leiva et al. (2015e).

5.4 PROBABILITY PLOTS

Notations and transformations that are useful for goodness of fit based on probability plots are provided next. In addition, we establish the hypotheses of interest and the corresponding test statistics for assessing goodness of fit in the Birnbaum–Saunders distribution with censored and uncensored data.

Let $T_1, T_2, \ldots, T_n$ be a sample of size n extracted from a random variable T following a distribution with cumulative distribution function $F_T(\cdot)$. Also, let $T_{1:n} \leq T_{2:n} \leq \cdots \leq T_{n:n}$ be the order statistics of $T_1, T_2, \ldots, T_n$,

Table 5.1 Interpretation of $2\log(B_{12})$ associated with the Bayes factor

$2\log(B_{12})$	Evidence in favor of M_1
< 0	Negative (M_2 is accepted)
$[0, 2)$	Weak
$[2, 6)$	Positive
$[6, 10)$	Strong
≥ 10	Very strong

with $t_{1:n} \leq t_{2:n} \leq \cdots \leq t_{n:n}$ and $t_1, t_2, \ldots, t_n$ being their corresponding observed values. The empirical cumulative distribution function of the data $t_1, t_2, \ldots, t_n$ is defined by

$$F_n(t) = \begin{cases} 0, & \text{if } t < t_{1:n}; \\ \frac{j}{n} = w_{j:n} + \frac{1}{2n}, & \text{if } t_{j:n} \leq t < t_{j+1:n}; \\ 1, & \text{if } t \geq t_{n:n}; \end{cases} \tag{5.4}$$

where $w_{j:n} = (2j - 1)/(2n)$, for $j = 1, \ldots, n - 1$.

As it is well-known, the random variable U given by the transformation

$$U = F_T(T) \tag{5.5}$$

follows a uniform distribution in the interval $[0, 1]$, which is denoted by $\mathrm{U}(0, 1)$, for any continuous theoretical cumulative distribution function $F_T(\cdot)$. The transformation defined in Equation (5.5) is known as the probability integral transform.

Another transformation associated with the $\mathrm{U}(0, 1)$ distribution was proposed by Michael (1983) and it is known as the stabilized probability (SP) transform. He noted that, if $U \sim \mathrm{U}(0, 1)$, then the random variable given by the transformation

$$S = \frac{2}{\pi} \arcsin\left(\sqrt{U}\right) \tag{5.6}$$

follows a distribution with probability density function

$$f_S(s) = \frac{2}{\pi} \sin(\pi s), \quad 0 < s < 1. \tag{5.7}$$

The order statistics $S_{1:n} \leq S_{2:n} \leq \cdots \leq S_{n:n}$ associated with a sample of size n from the distribution of the random variable S given in Equation (5.6) have a constant asymptotic variance, because as n goes to ∞ and j/n to q, $\mathrm{Var}(n\, S_{j:n})$ goes to $1/\pi^2$, which is independent of q, for $j = 1, \ldots, n$; see Michael (1983).

Consider a sample with censoring proportion p, which conducts to r failure (uncensored) data and $n - r$ censored data. Note that n and p are controlled by the researcher. Assume a type-II right censorship, so that r is fixed and $n - r$ observations are greater than the censoring point $t_{r:n}$. Then,

$$U_{1:n} = F_T(T_{1:n}) \leq U_{2:n} = F_T(T_{2:n}) \leq \cdots \leq U_{r:n} = F_T(T_{r:n})$$

are the smallest r order statistics of the type-II censored sample of size n and $n - r$ observations are greater than $u_{r:n} = F_T(t_{r:n})$. Let $U_{j:n} = F_T(T_{j:n})$ be the jth order statistic of a sample of size n extracted from a random variable

$U \sim \mathrm{U}(0,1)$ and $u_{j:n} = F_T(t_{j:n})$ its observed value, for $j = 1, \ldots, n$. Consider the transformation

$$U^*_{j:n} = \frac{U_{j:n}(B_{r,n-r+1}(U_{r:n}))^{1/r}}{U_{r:n}}, \tag{5.8}$$

for $j = 1, \ldots, r, r = 1, \ldots, n$, where $B_{r,n-r+1}(x) = I_x(r, n-r+1)$ is the cumulative distribution function of the beta distribution with parameters r and $n-r+1$ and $I_x(\cdot,\cdot)$ being the incomplete beta ratio function given by

$$I_x(a,b) = \frac{\beta(x;\, a,b)}{\beta(a,b)},$$

where $\beta(a,b)$ is the beta function, also called the Euler integral of the first kind, defined as

$$\beta(a,b) = \int_0^1 t^{a-1}(1-t)^{b-1}\, \mathrm{d}t\text{"}$$

and $\beta(x;a,b)$ is a generalization of the beta function given by

$$\beta(x;\, a,b) = \int_0^x t^{a-1}\,(1-t)^{b-1}\, \mathrm{d}t.$$

The order statistics $U^*_{1:n}, U^*_{2:n}, \ldots, U^*_{r:n}$ obtained from the transformation given in Equation (5.8) are distributed as the order statistics from a complete sample of size r from $U^* \sim \mathrm{U}(0,1)$; see details in Theorem 1 of Lin et al. (2008), Theorem 1 of Michael and Schucany (1979), and Theorem 8 of Fischer and Kamps (2011); see also Barros et al. (2014).

Consider the hypotheses:

H_0: The data come from a random variable T with cumulative distribution function $F_T(\cdot)$

versus (5.9)

H_1: The data do not come from this random variable.

The hypothesized distribution with cumulative distribution function $F_T(\cdot)$ is indexed by a parameter vector $\boldsymbol{\theta}$ that can contain location (μ), scale (β), and shape (α) parameters, or any other statistical parameter. We assume T follows the Birnbaum–Saunders distribution with shape (α) and scale (β) parameters, that is, $T \sim \mathrm{BS}(\alpha, \beta)$. If the cumulative distribution function is completely specified in H_0, that is, $\boldsymbol{\theta}$ is assumed to be known, the data must be transformed for testing uniformity. On the contrary, the parameters must be consistently estimated and the data transformed for testing normality from the distribution under H_0.

To test H_0 established in Equation (5.9), when $F_T(\cdot)$ is completely specified, and then to assess goodness of fit for the Birnbaum–Saunders distribution to a censored or uncensored data set, we consider test statistics based on the empirical cumulative distribution function $F_n(\cdot)$ defined in Equation (5.4). The most common statistics constructed with the empirical cumulative distribution function use vertical distances over t between $F_n(t)$ and $F_T(t)$ by means of the supremum and quadratic classes. Statistics that consider the mentioned classes are Anderson–Darling (AD), Cramér–von Mises (CM), KS, Kuiper (KU), and Watson (WA), which are given by

$$\mathrm{AD} = n\int_{-\infty}^{\infty} \frac{(F_n(t) - F_T(t))^2}{F_T(t)(1 - F_T(t))}\,\mathrm{d}F_T(t), \tag{5.10}$$

$$\mathrm{CM} = n\int_{-\infty}^{\infty} (F_n(t) - F_T(t))^2\,\mathrm{d}F_T(t), \tag{5.11}$$

$$\begin{aligned}\mathrm{KS} &= \sup_t |F_n(t) - F_T(t)| \\ &= \max\left\{\sup_t \{F_n(t) - F_T(t)\}, \sup_t \{F_T(t) - F_n(t)\}\right\},\end{aligned} \tag{5.12}$$

$$\mathrm{KU} = \sup_t \{F_n(t) - F_T(t)\} + \sup_t \{F_T(t) - F_n(t)\}, \tag{5.13}$$

$$\mathrm{WA} = n\int_{-\infty}^{\infty}\left(F_n(t) - F_T(t) - \int_{-\infty}^{\infty}(F_n(t) - F_T(t))\,\mathrm{d}F_T(t)\right)^2 \mathrm{d}F_T(t). \tag{5.14}$$

By considering the SP transform given in Equation (5.6) and the KS statistic given in Equation (5.12), we can define the Michael (MI) statistic as

$$\begin{aligned}\mathrm{MI} = \max\Bigg\{&\sup_t\left\{\frac{2}{\pi}\arcsin(F_n(t)) - \frac{2}{\pi}\arcsin(F_T(t))\right\}, \\ &\sup_t\left\{\frac{2}{\pi}\arcsin(F_T(t)) - \frac{2}{\pi}\arcsin(F_n(t))\right\}\Bigg\}.\end{aligned} \tag{5.15}$$

Now, by considering the probability integral transform given in Equation (5.5), AD, CM, KS, KU, MI, and WA statistics defined in Equations (5.10)–(5.15) can be implemented in practice by the formulas

$$\mathrm{AD} = -2\sum_{j=1}^{n}\left(w_{j:n}\log(U_{j:n}) + (1 - w_{j:n})\log(1 - U_{j:n})\right) - n, \tag{5.16}$$

$$\mathrm{CM} = \sum_{j=1}^{n}\left(U_{j:n} - w_{j:n}\right)^2 + \frac{1}{12n}, \tag{5.17}$$

$$\mathrm{KS} = \max\left\{\max_{1\le j\le n}\left\{w_{j:n} + \frac{1}{2n} - U_{j:n}\right\}, \max_{1\le j\le n}\left\{U_{j:n} - w_{j:n} + \frac{1}{2n}\right\}\right\}, \tag{5.18}$$

$$\mathrm{KU} = \max_{1\leq j\leq n}\left\{w_{j:n} + \frac{1}{2n} - U_{j:n}\right\} + \max_{1\leq j\leq n}\left\{U_{j:n} - w_{j:n} + \frac{1}{2n}\right\}, \tag{5.19}$$

$$\mathrm{MI} = \max\left\{\max_{1\leq j\leq n}\left(\frac{2}{\pi}\arcsin\left(w_{j:n} + \frac{1}{2n}\right) - \frac{2}{\pi}\arcsin(U_{j:n})\right),\right.$$
$$\left.\max_{1\leq j\leq n}\left\{\frac{2}{\pi}\arcsin(U_{j:n}) - \frac{2}{\pi}\arcsin\left(w_{j:n} - \frac{1}{2n}\right)\right\}\right\}, \tag{5.20}$$

$$\mathrm{WA} = \sum_{j=1}^{n}\left(U_{j:n} - w_{j:n}\right)^2 - n\left(\frac{1}{n}\sum_{j=1}^{n}U_{j:n} - \frac{1}{2}\right)^2 + \frac{1}{12n}, \tag{5.21}$$

where $w_{j:n}$ and $U_{j:n}$ are as given in Equations (5.4) and (5.8), respectively. For more details about expressions provided in Equations (5.10) through (5.21), the interested reader is referred to Michael (1983) and D'Agostino and Stephens (1986, Ch. 4).

Quantiles of the distribution of the statistics AD, CM, KS, KU, MI, and WA must be obtained under H_0. However, if the distribution under this hypothesis is not completely specified, its parameters must be properly estimated and the AD, CM, KS, KU, MI, and WA statistics must be modified for the distribution under H_0. These modified statistics are denoted by AD*, CM*, KS*, KU*, MI*, and WA*, and their calculated values by ad*, cm*, ks*, ku*, mi*, and wa*, respectively. In this case, new quantiles of the distributions of AD*, CM*, KS*, KU*, MI*, and WA* must be computed under H_0.

When a parametric statistical analysis with censored data needs to validate its distributional assumption, the classic goodness-of-fit test statistics need to be adapted for considering the censorship in one of two options. The first of them consists of using goodness-of-fit tests for uncensored data adapting the type-II right censored data to become an uncensored (complete) data sample. The second option allows us to adapt the test statistics to type-II right censored data; see, for example, Malmquist (1950), D'Agostino and Stephens (1986, Chs. 4 and 11), Lin et al. (2008), and Barros et al. (2014).

To test H_0 when $F_T(\cdot)$ is completely specified and then to assess goodness of fit in practice with r uncensored data and $n-r$ type-II right censored data, we use the results presented in D'Agostino and Stephens (1986, Ch. 4) and adapt the statistics given in Equations (5.16)–(5.21) as

$$\mathrm{AD}_{r,n} = -2\sum_{j=1}^{r}\left(w_{j:n}\log(U_{j:n}) - (1 - w_{j:n})\log(1 - U_{j:n})\right)$$

$$-\frac{1}{n}\left((n-r)^2\log(1-U_{r:n})-r^2\log(U_{r:n})+n^2U_{r:n}\right), \quad (5.22)$$

$$\mathrm{CM}_{r,n}=\sum_{j=1}^{r}(U_{j:n}-w_{j:n})^2+\frac{r}{12n^2}+\frac{n}{3}\left(U_{r:n}-\frac{r}{n}\right)^3, \quad (5.23)$$

$$\mathrm{KS}_{r,n}=\max\left\{\max_{1\le j\le r}\left\{w_{j:n}+\frac{1}{2n}-U_{j:n}\right\},\max_{1\le j\le r}\left\{U_{j:n}-w_{j:n}+\frac{1}{2n}\right\}\right\}, \quad (5.24)$$

$$\mathrm{KU}_{r,n}=\max_{1\le j\le r}\left\{w_{j:n}+\frac{1}{2n}-U_{j:n}\right\}+\max_{1\le j\le r}\left\{U_{j:n}-w_{j:n}+\frac{1}{2n}\right\}, \quad (5.25)$$

$$\mathrm{MI}_{r,n}=\max\left\{\max_{1\le j\le r}\left\{\frac{2}{\pi}\arcsin\left(w_{j:n}+\frac{1}{2n}\right)-\frac{2}{\pi}\arcsin(U_{j:n})\right\},\right.$$
$$\left.\max_{1\le j\le r}\left\{\frac{2}{\pi}\arcsin(U_{j:n})-\frac{2}{\pi}\arcsin\left(w_{j:n}-\frac{1}{2n}\right)\right\}\right\}, \quad (5.26)$$

$$\mathrm{WA}_{r,n}=\sum_{j=1}^{r}(U_{j:n}-w_{j:n})^2+\frac{r}{12n^2}+\frac{n}{3}\left(U_{r:n}-\frac{r}{n}\right)^3$$
$$-nU_{r:n}\left(\frac{r}{n}-\frac{U_{r:n}}{2}-\frac{\sum_{j=1}^{r}U_{j:n}}{nU_{r:n}}\right)^2. \quad (5.27)$$

The quantiles of the distributions of the $\mathrm{AD}_{r,n}$, $\mathrm{CM}_{r,n}$, $\mathrm{KS}_{r,n}$, $\mathrm{KU}_{r,n}$, $\mathrm{MI}_{r,n}$, and $\mathrm{WA}_{r,n}$ statistics given in Equations (5.22)–(5.27) must be obtained under H_0. However, as mentioned in the uncensored case, if the distribution under H_0 is not completely specified, its parameters must be properly estimated, taking into account the censorship, and the statistics must be modified for each case under H_0. We denote these statistics by $\mathrm{AD}^*_{r,n}$, $\mathrm{CM}^*_{r,n}$, $\mathrm{KS}^*_{r,n}$, $\mathrm{KU}^*_{r,n}$, $\mathrm{MI}^*_{r,n}$, and $\mathrm{WA}^*_{r,n}$, and their calculated values by $\mathrm{ad}^*_{r,n}$, $\mathrm{cm}^*_{r,n}$, $\mathrm{ks}^*_{r,n}$, $\mathrm{ku}^*_{r,n}$, $\mathrm{mi}^*_{r,n}$, and $\mathrm{wa}^*_{r,n}$, respectively. Also, new quantiles of the distributions of $\mathrm{AD}^*_{r,n}$, $\mathrm{CM}^*_{r,n}$, $\mathrm{KS}^*_{r,n}$, $\mathrm{KU}^*_{r,n}$, $\mathrm{MI}^*_{r,n}$, and $\mathrm{WA}^*_{r,n}$ must be computed under H_0. For more details about how to obtain the quantiles of the distributions of the corresponding test statistics under H_0, which have been studied for different distributions of the location-scale family with uncensored and censored, see D'Agostino and Stephens (1986), Castro-Kuriss et al. (2009, 2010), and Castro-Kuriss (2011). For the Birnbaum–Saunders distribution, analogous results for assessing goodness of fit with both uncensored and censored data can be considered, which we detail below.

We consider a procedure that can be applied to the Birnbaum–Saunders distribution based on the work proposed by Chen and Balakrishnan (1995), which provides an approximate goodness-of-fit method for this distribution.

The method first transforms the data to normality and then applies the algorithm presented in Chen and Balakrishnan (1995), generalizing it. Testing normality in H_0 allows us to compute the critical values of the corresponding test statistics, independently of the parameter estimators, if they are consistent and the sample size is large enough. To test the hypotheses of interest defined in Equation (5.9) with $T \sim \mathrm{BS}(\alpha, \beta)$, for $\alpha > 0$ and $\beta > 0$ unknown, we consider a generalization of the algorithm presented in Chen and Balakrishnan (1995), which is detailed in Algorithm 5. Following these authors, we recommend in general to use a sample size $n > 20$, so that the approximations work well. This is also valid for the algorithms presented below.

Algorithm 5 Goodness-of-fit test for the Birnbaum–Saunders distribution with uncensored data

1: Collect data $t_1, t_2, \ldots, t_n$ and order them as $t_{1:n}, t_{2:n}, \ldots, t_{n:n}$;
2: Estimate α and β of $F_T(t; \alpha, \beta)$ given in Equation (2.7) by $\hat{\alpha}$ and $\hat{\beta}$, respectively, with $t_1, t_2, \ldots, t_n$;
3: Compute $\hat{v}_{j:n} = F_T(t_{j:n}; \hat{\alpha}, \hat{\beta})$, for $j = 1, \ldots, n$;
4: Calculate $\hat{y}_j = \Phi^{-1}(\hat{v}_{j:n})$, where $\Phi^{-1}(\cdot)$ is the inverse function of the N(0, 1) cumulative distribution function $\Phi(\cdot)$;
5: Obtain $\hat{u}_{j:n} = \Phi(\hat{z}_j)$, where

$$\hat{z}_j = \frac{\hat{y}_j - \bar{y}}{s_y},$$

with

$$\bar{y} = \frac{1}{n}\sum_{j=1}^{n} \hat{y}_j, \quad s_y^2 = \frac{1}{n-1}\sum_{j=1}^{n} (\hat{y}_j - \bar{y})^2;$$

6: Evaluate AD*, CM*, KS*, KU*, MI*, and WA* statistics at $\hat{u}_{j:n}$;
7: Compute the p-values of the AD*, CM*, KS*, KU*, MI*, and WA* statistics;
8: Reject H_0: $T \sim \mathrm{BS}(\alpha, \beta)$ with $F_T(t; \alpha, \beta)$ for a specified significance level based on the obtained p-values.

As mentioned, goodness-of-fit tests for the Birnbaum–Saunders distribution with uncensored data can be considered for censored data adapting them or the goodness-of-fit statistics. To test the hypotheses of interest defined in Equation (5.9) with $T \sim \mathrm{BS}(\alpha, \beta)$, for $\alpha > 0$ and $\beta > 0$,

Algorithm 6 Goodness-of-fit test 1 for the Birnbaum–Saunders distribution with censored data

1: Collect data $t_1, t_2, \ldots, t_n$ and order them as $t_{1:n}, t_{2:n}, \ldots, t_{n:n}$;
2: Estimate α and β of $F_T(t; \alpha, \beta)$ given in Equation (2.7) by $\hat{\alpha}$ and $\hat{\beta}$, respectively, with censored data $t_1, t_2, \ldots, t_n$;
3: Compute $\hat{v}_{j:n} = F_T(t_{j:n}; \hat{\alpha}, \hat{\beta})$, for $j = 1, \ldots, n$;
4: Determine

$$\hat{v}^*_{j:n} = \frac{\hat{v}_{j:n}(B_{r,n-r+1}(\hat{v}_{r:n}))^{1/r}}{\hat{v}_{r:n}},$$

for $j = 1, \ldots, r$ and $r = 1, \ldots, n$;
5: Calculate $\hat{y}_j = \Phi^{-1}(\hat{v}^*_{j:n})$;
6: Obtain $\hat{u}_{j:n} = \Phi(\hat{z}_j)$, where $\hat{z}_j = (\hat{y}_j - \bar{y})/s_y$, with $\bar{y}, s_y$ given in Step 5 of Algorithm 5;
7: Evaluate AD*, CM*, KS*, KU*, MI*, and WA* statistics at $\hat{u}_{j:n}$;
8: Compute the p-values of the AD*, CM*, KS*, KU*, MI*, and WA* statistics;
9: Reject H$_0$: $T \sim \text{BS}(\alpha, \beta)$ with $F_T(t; \alpha, \beta)$ for a specified significance level based on the obtained p-values.

both of them unknown and type-II right censored data, we first transform censored data into uncensored data by using the transformation given in Equation (5.8). Algorithm 6 details the corresponding goodness-of-fit procedure. Second, as mentioned, another way to perform a goodness-of-fit test for the Birnbaum–Saunders distribution with censored data can be obtained adapting the goodness-of-fit statistics, which is detailed in Algorithm 7.

PP and QQ plots are well known, unlike the SP plot. We recall that, if the distribution under H$_0$ is U(0, 1), then the corresponding QQ plot is essentially the same as the PP plot; see Castro-Kuriss et al. (2009). As mentioned, Michael (1983) used the arcsin transformation given in Equation (5.6) to stabilize the variance of the points on probability graphs associated with the KS test and to propose the SP plot. Formulas to construct PP and SP plots are provided in Table 5.2. In this table, $w_{j:n}$ is as given in Equation (5.4) and $u_{j:n}$ as in Equation (5.8).

Acceptance regions for PP and SP plots can be constructed by means of KS and MI statistics. Thus, we can display acceptance bands to assess whether the data can come from the distribution under H$_0$ with these

Algorithm 7 Goodness-of-fit test 2 for the Birnbaum–Saunders distribution with censored data

1: Collect data $t_1, t_2, \ldots, t_n$ and order them as $t_{1:n}, t_{2:n}, \ldots, t_{n:n}$;
2: Estimate α and β of $F_T(t; \alpha, \beta)$ given in Equation (2.7) by $\hat{\alpha}$ and $\hat{\beta}$, respectively, with censored data $t_1, t_2, \ldots, t_n$;
3: Compute $\hat{v}_{j:n} = F_T(t_{j:n}; \hat{\alpha}, \hat{\beta})$, for $j = 1, \ldots, n$;
4: Calculate $\hat{y}_j = \Phi^{-1}(\hat{v}_{j:n})$;
5: Obtain $\hat{u}_{j:n} = \Phi(\hat{z}_j)$, where $\hat{z}_j = (\hat{y}_j - \bar{y})/s_y$;
6: Evaluate $\text{AD}^*_{r,n}$, $\text{CM}^*_{r,n}$, $\text{KS}^*_{r,n}$, $\text{KU}^*_{r,n}$, $\text{MI}^*_{r,n}$, and $\text{WA}^*_{r,n}$ statistics at $\hat{u}_{j:n}$;
7: Determine the p-values of $\text{AD}^*_{r,n}$, $\text{CM}^*_{r,n}$, $\text{KS}^*_{r,n}$, $\text{KU}^*_{r,n}$, $\text{MI}^*_{r,n}$, and $\text{WA}^*_{r,n}$ statistics;
8: Reject the corresponding H_0 for a specified significance level depending on the obtained p-values.

two statistics; see Castro-Kuriss et al. (2009, 2010). Formulas to construct $100\varrho\%$ acceptance regions on PP and SP plots with right type-II censored data, based on $\text{KS}^*_{r,n}$ and $\text{MI}^*_{r,n}$ statistics, are displayed in Table 5.3. In this table, w and x are continuous versions of $w_{j:n}$ and $x_{j:n}$ given in Table 5.2 to construct the acceptance bands. If all of the r data points lie inside the

Table 5.2 Formulas for the indicated probability plot

Plot	Abscissa	Ordinate
PP	$w_{j:n}$	$u_{j:n}$
SP	$x_{j:n} = \frac{2}{\pi}\arcsin\left(\sqrt{w_{j:n}}\right)$	$s_{j:n} = \frac{2}{\pi}\arcsin\left(\sqrt{u_{j:n}}\right)$

Table 5.3 $100\varrho\%$ acceptance regions for the indicated plot and statistic with 100ϱth quantiles $\text{ks}^*_{r,n,\varrho}$ and $\text{mi}^*_{r,n,\varrho}$

Plot	Statistic	Bands defining acceptance regions
PP	KS*	$(\max\{w - \text{ks}^*_{r,n,\varrho} + \frac{1}{2n}, 0\}, \min\{w + \text{ks}^*_{r,n,\varrho} - \frac{1}{2n}, 1\})$
PP	MI*	$(\max\{\sin^2(\arcsin(w^{1/2}) - \frac{\pi}{2}\text{mi}^*_{r,n,\varrho}), 0\},$ $\min\{\sin^2(\arcsin(w^{1/2}) + \frac{\pi}{2}\text{mi}^*_{r,n,\varrho}), 1\})$
SP	KS*	$(\max\{\frac{2}{\pi}\arcsin(\{\sin^2(\frac{\pi}{2}x) - \text{ks}^*_{r,n,\varrho} + \frac{1}{2n}\}^{1/2}), 0\},$ $\min\{\frac{2}{\pi}\arcsin(\{\sin^2(\frac{\pi}{2}x) + \text{ks}^*_{r,n,\varrho} - \frac{1}{2n}\}^{1/2}), 1\})$
SP	MI*	$(\max\{x - \text{mi}^*_{r,n,\varrho}, 0\}, \min\{x + \text{mi}^*_{r,n,\varrho}, 1\})$

constructed acceptance bands, then H_0 cannot be rejected at the $1-\varrho$ level. Also, if a noticeable curvature is detected, we can question such a hypothesis. Table 5.3 may be adapted to the uncensored case with $r=n$ and the quantiles must be replaced by the quantiles of the distribution of the corresponding statistics without censorship.

To test H_0 defined in Equation (5.9) with $T \sim \mathrm{BS}(\alpha, \beta)$, for some $\alpha > 0$ and $\beta > 0$ and type-II right censored data, we consider a goodness-of-fit graphical tool whose procedure is detailed in Algorithm 8 for PP and SP plots. This tool is valid for censored or uncensored data and is based on Algorithm 7 and Tables 5.2–5.3. We consider the general case for unknown parameters of the distribution under H_0, but it can also be used when the parameters are known. The interested reader is referred to Castro-Kuriss et al. (2014) for more details about goodness-of-fir graphical tools based on KS and MI tests.

Algorithm 8 Acceptance regions to test goodness of fit in the Birnbaum–Saunders distribution with censored data

1: Collect data $t_1, t_2, \ldots, t_n$ and order them as $t_{1:n}, t_{2:n}, \ldots, t_{n:n}$;
2: Estimate α and β of $F_T(t; \alpha, \beta)$ given in Equation (2.7) by $\hat{\alpha}$ and $\hat{\beta}$, respectively, with censored data $t_1, t_2, \ldots, t_n$;
3: Compute $\hat{v}_{j:n} = F_T(t_{j:n}; \hat{\alpha}, \hat{\beta})$, for $j = 1, \ldots, n$;
4: Calculate $\hat{y}_j = \Phi^{-1}(\hat{v}_{j:n})$;
5: Obtain $\hat{u}_{j:n} = \Phi(\hat{z}_j)$, where $\hat{z}_j = (\hat{y}_j - \bar{y})/s_y$;
6: Draw the PP plot with points $w_{j:n}$ computed as in Equation (5.4) versus points $\hat{u}_{j:n}$, for $j = 1, \ldots, r$ and $r = 1, \ldots, n$;
7: Draw the SP plot with points

$$x_{j:n} = \frac{2}{\pi} \arcsin\left(\sqrt{w_{j:n}}\right) \quad \text{versus} \quad s_{j:n} = \frac{2}{\pi} \arcsin\left(\sqrt{\hat{u}_{j:n}}\right);$$

8: Construct acceptance bands according to Table 5.3 specifying an $1-\varrho$ significance level;
9: Decide if H_0 must be rejected for the specified significance level;
10: Corroborate decision in Step 9 with the p-values after evaluating $\mathrm{KS}^*_{r,n}$ and $\mathrm{MI}^*_{r,n}$ statistics at $\hat{u}_{j:n}$.

CHAPTER 6

Data Analyses with the Birnbaum–Saunders Distribution

Abstract

In this chapter, several data sets which have been modeled by the Birnbaum–Saunders distribution are provided. We describe these data sets for their potential use by means of the tools described in Chapters 2–5 for the Birnbaum–Saunders distribution. Two existing packages of the R software created for this distribution, called bs and gbs, are discussed by means of the analysis of a data set, which can be easily replicated to the other data sets. This chapter is finished showing how the statistical modeling and diagnostics presented in Chapter 4 can be carried out with some of these data sets.

Keywords: Data analysis, diagnostic methods, goodness-of-fit tools, maximum likelihood estimation, regression models.

6.1 INTRODUCTION

We provide several data sets which have been modeled by the Birnbaum–Saunders distribution. These data sets are divided in two groups: (i) fatigue databases for different types of materials, such as aluminum, concrete, Kevlar fiber, metal, steel, and wood, which should be well modeled by the Birnbaum–Saunders distribution due to its physical arguments; and (ii) lifetime databases for different types of specimens from engineering, such as bearing, insulating fluid, machine valves, transceivers, and welds, as well as from industry and medicine, such as balls for electronic use, food products, patients, and pigs, which also are related in someway to the genesis of the Birnbaum–Saunders distribution.

The Birnbaum–Saunders Distribution. http://dx.doi.org/10.1016/B978-0-12-803769-0.00006-6

6.2 DATA SETS

Fatigue-life data

Aluminum

First, Birnbaum and Saunders (1958) collected fatigue-life data corresponding to cycles ($\times 10^{-3}$) until failure of aluminum specimens of type 6061-T6. The specimens were cut parallel to the direction of rolling and oscillating at 18 cycles per second. These specimens were exposed to stress levels of 21,000 (S1), 26,000 (S2), and 31,000 (S3) pounds per square inch (psi) for $n = 101, 102, 101$ specimens for each stress level, respectively. The data sets are displayed in Table 6.1; see also Leiva et al. (2014b). Second, Collins (1981) provided another data set of size $n = 35$ related to fatigue-life (in cycles $\times 10^{-3}$) of aluminum specimens, which were tested until failure at a stress level of 26,000 (S4) psi. This data set is displayed in Table 6.1; see also Mills (1997).

Concrete

These data sets correspond to the fatigue-life (in cycles $\times 10^{-3}$) of concrete specimens. Each set is characterized by the ratio 0.95, 0.90, and 0.825 of applied stress causing failure. The number of cycles until failure is expected to increase inversely with this ratio. Three different data sets each of $n = 15$ observations are provided in Table 6.2; see Holmen (1979) and Mills (1997).

Kevlar fiber

These data were analyzed by Smith (1991) and correspond to failure times (measured in hours) of a particular kind of fiber (named Kevlar 49) subject to different stress levels (measured in MPa). A random sample of eight spools was obtained from the population of spools. Then, each spool was subject to different stress levels in $n = 108$ specimens of fiber until failure or censoring at 41,000 h using type I censoring. The data are displayed in Table 6.3; see also Villegas et al. (2011).

Metal

First, data corresponding to biaxial fatigue-life of $n = 46$ metal specimens (in cycles) until failure are provided. This data set is shown in Table 6.4; see Rieck (1989). Second, a data set also related to biaxial fatigue reported by Brown and Miller (1978) on the fatigue-life of metal specimens until failure (response variable) and one explanatory variable are considered. The response variable (T) is the number of cycles to failure and the explanatory variable or covariable (X) is the work per cycle in MJ/m^3. A number

Table 6.1 Fatigue life (in cycles × 10^{-3}) of aluminum specimens exposed to the indicated stress level

21,000 psi (S1)					26,000 psi (S2)					31,000 psi (S3)					26,000 psi (S4)	
370	706	716	746	785	233	258	268	276	290	70	90	96	97	99	233	410
797	844	855	858	886	310	312	315	318	321	100	103	104	104	105	276	422
886	930	960	988	999	321	329	335	336	338	107	108	108	108	109	290	433
1000	1010	1016	1018	1020	338	342	342	342	344	109	112	112	113	114	310	439
1055	1085	1102	1102	1108	349	350	350	351	351	114	114	116	119	120	315	439
1115	1120	1134	1140	1199	352	352	356	358	358	120	120	121	121	123	321	443
1200	1200	1203	1222	1235	360	362	363	366	367	124	124	124	124	124	342	445
1238	1252	1258	1262	1269	370	370	372	372	374	128	128	129	129	130	351	445
1270	1290	1293	1300	1310	375	376	379	379	380	130	130	131	131	131	356	456
1313	1315	1330	1355	1390	382	389	389	395	396	131	131	132	132	132	358	473
1416	1419	1420	1420	1450	400	400	400	403	404	133	134	134	134	134	358	490
1452	1475	1478	1481	1485	406	408	408	410	412	134	136	136	137	138	360	517
1502	1505	1513	1522	1522	414	416	416	416	420	138	138	139	139	141	362	540
1530	1540	1560	1567	1578	422	423	426	428	432	141	142	142	142	142	367	560
1594	1602	1604	1608	1630	432	433	433	437	438	142	142	144	144	145	376	
1642	1674	1730	1750	1750	439	439	443	445	445	146	148	148	149	151	395	
1763	1768	1781	1782	1792	452	456	456	460	464	151	152	155	156	157	400	
1820	1868	1881	1890	1893	466	468	470	470	473	157	157	157	158	159	400	
1895	1910	1923	1924	1945	474	476	476	486	488	162	163	163	164	166	403	
2023	2100	2130	2215	2268	489	490	491	503	517	166	168	170	174	196	404	
2440					540	560				212					406	

Table 6.2 Fatigue-life (in cycles $\times 10^{-3}$) of concrete specimens exposed to the indicated ratio

Ratio	Fatigue-life														
0.95	37	72	74	76	83	85	105	109	120	123	143	203	206	217	257
0.90	201	216	226	252	257	295	311	342	356	451	457	509	540	680	1129
0.825	1246	1258	1460	1492	2400	2410	2590	2903	3330	3590	3847	4110	4820	5560	5598

Table 6.3 Failure time (in hours) and censoring indicator (1: censoring; 0: noncensoring) of Kevlar 49 fiber specimens for the indicated spool and stress level (measured in MPa)

Spool	Stress level	(Failure time, censoring indicator)
1	23.4	(41,000, 0); (41,000, 0); (41,000, 0); (41,000, 0);
	25.5	(11,487.3, 1); (14,032.0, 1); (31,008.0, 1);
	27.6	(453.4, 1); (664.5, 1); (930.4, 1); (1755.5, 1);
	29.7	(444.4, 1); (755.2, 1); (952.2, 1); (1108.2, 1);
2	23.4	(14,400, 1);
	25.5	(1134.3, 1); (1824.3, 1); (1920.1, 1); (2383.0, 1); (3708.9, 1); (5556.0, 1);
	27.6	(71.2, 1); (199.1, 1); (403.7, 1); (432.2, 1); (514.1, 1); (544.9, 1); (694.1, 1);
	29.7	(2.2, 1); (8.5, 1); (9.1, 1); (10.2, 1); (22.1, 1); (55.4, 1); (111.4, 1); (158.7, 1);
3	23.4	(8616, 1);
	25.5	(1087.7, 1); (2442.5, 1);
	27.6	(19.1, 1); (24.3, 1); (69.8, 1); (136.0, 1);
	29.7	(12.5, 1); (14.6, 1); (18.7, 1); (101.0, 1);
4	23.4	(41,000, 0); (41,000, 0); (41,000, 0); (41,000, 0);
	25.5	(13,501.3, 1); (29,808.0, 1);
	27.6	(876.7, 1); (1275.6, 1); (1536.8, 1); (6177.5, 1);
	29.7	(254.1, 1); (1148.5, 1); (1569.3, 1); (1750.6, 1); (1802.1, 1);
5	23.4	(9120, 1); (20,231, 1); (35,880, 1);
	25.5	(11,727.1, 1)
	27.6	–
	29.7	(8.3, 1); (13.3, 1); (87.5, 1); (243.9, 1);
6	23.4	(7320, 1); (16,104, 1); (20,233, 1);
	25.5	(225.2, 1); (6271.1, 1); (7996.0, 1);
	27.6	(514.2, 1); (541.6, 1); (1254.9, 1);
	29.7	(6.7, 1); (15.0, 1); (144.0, 1);
7	23.4	(4000, 1); (5376, 1);
	25.5	(503.6, 1);
	27.6	–
	29.7	(4.0, 1); (4.0, 1); (4.6, 1); (6.1, 1); (7.9, 1); (14.0, 1); (45.9, 1); (61.2, 1);
8	23.4	(41,000, 0); (41,000, 0); (41,000, 0);
	25.5	(2974.6 1); (4908.9 1); (7332.0 1); (7918.7 1); (9240.3 1); (9973.0 1);
	27.6	(554.2, 1); (2046.2, 1);
	29.7	(98.2, 1); (590.4, 1); (638.2, 1).

Table 6.4 Biaxial fatigue life of metal specimens

125	127	135	137	185	187	190	190	195	200	212	242	245	255	283	316
327	355	373	386	456	482	552	580	700	736	745	750	804	852	884	977
1040	1066	1093	1114	1125	1300	1536	1583	2208	2266	2834	3280	4707	5046		

Table 6.5 Biaxial fatigue life of metal specimens

X	T	X	T	X	T	X	T
11.5	3280	24.0	804	40.1	750	60.3	283
13.0	5046	24.6	1093	40.1	316	60.5	212
14.3	1563	25.2	1125	43.0	456	62.1	327
15.6	4707	25.5	884	44.1	552	62.8	373
16.0	977	26.3	1300	46.5	355	66.5	125
17.3	2834	27.9	852	47.3	242	67.0	187
19.3	2266	28.3	580	48.7	190	67.1	135
21.1	2208	28.4	1066	52.9	127	67.9	245
21.5	1040	28.6	1114	56.6	185	68.8	137
22.6	700	30.9	386	59.9	255	75.4	200
22.6	1583	31.9	745	60.2	195	100.5	190
24.0	482	34.5	736				

of $n = 46$ observations were considered. The data set is presented in Table 6.5. Third, die fatigue fracture is a typical metal fatigue caused by cyclic stress in the course of the service life cycle of dies (die lifetime). Although this fatigue could be mainly determined by die lifetime, other random variables can also be considered as responses to this fatigue. Then, it is of interest to model fatigue in a metal forming process for $n = 19$ observations considering as responses to Von Mises stress (T_1, in N/mm^2), maximum deformation (T_2, dimensionless), manufacturing force (T_3, in Newton [N]) and die lifetime (T_4, in number of cycles), whereas the covariables that could affect these responses are friction coefficient (X_1, dimensionless), angle of die (X_2, in °) and work temperature (X_3, in °C). The data were taken from Lepadatu et al. (2005) and are displayed in Table 6.6; see also Rieck and Nedelman (1991).

Steel

Three stress levels of 55,000, 61,000, and 67,000 psi were investigated to detect the fatigue life of steel specimens, all from a single heat of material.

Table 6.6 Responses of metal specimens for the indicated covariables

X_1	X_2	X_3	T_1	T_2	T_3	T_4
0.07	23.00	581.08	1850	1.260	144,000	6420
0.07	23.00	818.92	470	1.349	36,700	33,700
0.07	31.96	581.08	1830	1.532	156,000	9430
0.07	31.96	818.92	523	1.614	39,900	36,600
0.13	23.00	581.08	2030	1.801	181,000	12,100
0.13	23.00	818.92	581	1.824	46,900	32,000
0.13	31.96	581.08	2230	1.939	203,000	13,200
0.13	31.96	818.92	632	1.928	52,300	32,100
0.05	27.50	700.00	889	1.275	78,600	19,900
0.15	27.50	700.00	1410	1.921	125,000	15,000
0.10	20.00	700.00	1060	1.692	92,100	20,900
0.10	35.00	700.00	1390	1.888	111,000	21,200
0.10	27.50	500.00	2430	1.666	213,000	9170
0.10	27.50	900.00	243	1.685	19,500	74,800
0.10	27.50	700.00	1130	1.651	96,900	19,900

The corresponding fatigue-life data of $n = 20$ specimens are reported in cycles ($\times 10^{-3}$) in Table 6.7; see Collins (1981) and Mills (1997).

Wood

The data consist of the number of cycles corresponding to the fatigue life in wood specimens without amendments in two varieties, namely, *Pinus caribaea* Hondurenses and *Eucalyptus grandis*. Table 6.8 presents the data sets, where T is the response variable related to the total number of cycles until failure, which was studied taking into account $n = 6$ replicates (T_{ir}, for $r = 1, \ldots, 6$), X_1 is the level of stress (in %) and X_2 the level of frequency (in %). The data were obtained by the Laboratory of Wood and Timber Structures of the School of Engineering of São Carlos, University of São Paulo, Brazil; see Macedo (2000) and Martínez and Calil (2003).

Lifetimes data

Airborne transceivers

These data correspond to maintenance reported in von Alven (1964) on active repair-times (in hours) for $n = 46$ airborne communication transceivers, which are given in Table 6.9; see also Chhikara and Folks (1977).

Table 6.7 Fatigue life (in cycles $\times 10^{-3}$) of steel specimens for the indicated stress level

Stress level (psi)	Fatigue-life																			
55,000	174	236	242	257	271	295	319	352	357	377	415	438	458	493	552	578	685	696	1390	1676
61,000	78	82	85	86	95	98	101	102	105	105	123	126	126	127	137	147	148	158	162	175
67,000	30	31	31	32	33	33	38	39	41	44	45	45	45	46	46	46	50	54	64	75

Table 6.8 Fatigue-life (in cycles) of wood specimens (i) for the indicated specie and factor

Index i	T_{i1}	T_{i2}	T_{i3}	T_{i4}	T_{i5}	T_{i6}	X_1	X_2
Pinus caribaea Hondurensis								
1	211,480	430,899	204,017	229,270	252,510	400,197	60	1
2	93,432	118,610	178,388	66,818	72,917	87,134	75	1
3	91	57	21	139	113	87	90	1
4	900,005	703,389	1,794,615	1,149,067	1,382,568	791,034	60	5
5	360,056	124,050	309,413	163,539	253,120	219,832	75	5
6	187	417	348	617	584	381	90	5
7	793,417	1,106,942	1,814,582	1,264,336	1,330,109	1,016,584	60	9
8	150,375	241,314	454,026	264,474	297,954	373,929	75	9
9	103	325	649	619	669	925	90	9
Eucalyptus grandis								
1	300,060	447,727	256,633	483,367	428,444	180,865	60	1
2	153,472	124,896	81,872	94,734	77,639	114,332	75	1
3	13	45	204	104	34	134	90	1
4	1,601,572	902,906	1,310,678	867,411	1,413,229	1,887,691	60	5
5	339,606	255,226	356,979	249,317	182,182	290,385	75	5
6	181	120	236	639	594	897	90	5
7	963,383	1,367,100	1,410,236	1,145,760	1,145,760	1,535,709	60	9
8	175,454	241,199	310,731	473,653	332,578	329,510	75	9
9	355	157	430	284	745	1077	90	9

Table 6.9 Repair lifetimes (in hours) of airborne transceivers

0.2	0.3	0.5	0.5	0.5	0.5	0.6	0.6	0.7	0.7	0.7	0.8	0.8	1.0	1.0	1.0
1.0	1.1	1.3	1.5	1.5	1.5	1.5	2.0	2.0	2.2	2.5	2.7	3.0	3.0	3.3	3.3
4.0	4.0	4.5	4.7	5.0	5.4	5.4	7.0	7.5	8.8	9.0	10.3	22.0	24.5		

Table 6.10 Fatigue life of ball bearings

McCool											
152.7	172.0	172.5	173.3	193.0	204.7	216.5	234.9	262.6	422.6		
Lawless											
17.88	28.92	33.00	41.52	42.12	45.60	48.40	51.84	51.96	54.12	55.56	67.80
68.84	68.84	80.00*	80.00*	80.00*	80.00*	80.00*	80.00*	80.00*	80.00*	80.00*	

Ball bearing

McCool (1974) provided fatigue life (in hours) of $n = 10$ ball bearings of a certain type, which are displayed in Table 6.10; see also Ng et al. (2003). Lawless (1982) also presented data of failure times of $n = 23$ ball bearings with censoring, which are also displayed in Table 6.10, with "*" denoting a censored observation.

Ball size for electronic industry

Manufacturing of integrated circuits comprises the initial process of wafer and the final process of packaging. In an integrated circuit packaging factory, the manufacturing process generally includes: die sawing, die mounting, wire bonding, molding, trimming, forming, marking, plating, and testing. The wire bonding is the most common way to provide an electronic connection from the integrated circuit apparatus to the lead-frame. This is done by using an ultra-thin gold or aluminum wire to form the electrical interconnection between the chip and package leads. The high-speed wire bonding equipment consists of a control system to feed the lead frame toward the work area. The image recognition system guarantees that the die is oriented to match the bonding diagram for a particular device. The wires are bonded one at a time with two bonds for each connection: one in the die (first bond) and the other in the lead frame (second bond). The first bond requires a ball formation that is put within the bond pad opening on the die, under load and ultrasonic energy for a few milliseconds, forming a ball bond to the bond pad metal. In the wire bonding process, one of the most important factors that is directly related to the quality level is the ball size. Because the process can be interrupted and shut down when the width between the two bond balls is too small, the bond ball size must be considered. The data correspond to the size (in millimeters) of $n = 100$ balls and are displayed in Table 6.11; see Hsu et al. (2008) and Leiva et al. (2014a).

Fracture toughness of welds

This data set corresponds to the fracture toughness of $n = 19$ welds and is displayed in Table 6.12; see Ang and Tang (1975) and Folks and Chhikara (1978).

Guinea pigs

These data correspond to the survival times of guinea pigs injected with different doses of tubercle bacilli. It is well-known that guinea pigs have high susceptibility to human tuberculosis. This is the reason why they were used in the study. The researchers are primarily concerned with the animals

Table 6.11 Size (in millimeters) of balls for electronic industry

2.891	4.035	4.495	2.890	2.312	3.158	5.228	3.334	5.896	5.639
3.842	1.590	1.954	1.842	0.680	2.752	1.301	2.260	0.889	2.381
0.619	2.788	1.050	3.750	3.508	6.123	6.549	5.954	2.207	4.417
4.805	1.516	2.227	2.797	1.636	1.066	0.940	4.101	4.542	1.295
1.770	3.492	5.706	3.722	6.644	2.472	1.383	4.494	1.694	2.892
2.111	3.591	2.093	3.222	2.891	2.582	0.665	3.234	1.102	1.083
1.508	1.811	2.803	6.659	0.923	6.229	3.177	2.333	1.311	4.419
2.495	0.921	4.061	9.725	1.600	4.281	3.360	1.131	1.618	4.489
3.696	1.982	2.413	5.480	1.992	2.573	1.845	4.620	6.221	1.694
4.882	1.380	3.982	2.260	2.366	2.899	3.782	2.336	1.175	3.055

Table 6.12 Fracture toughness lifetime of weld specimens

54.4	62.6	63.2	67.0	70.2	70.5	70.6	71.4	71.8	74.1
74.1	74.3	78.8	81.8	83.0	84.4	85.3	86.9	87.3	

in the same cage that were under the same regimen. The regimen number is the common logarithm of the number of bacillary units in 0.5 mL of challenge solution, that is, regimen 6.6 corresponds to 4.0×106 bacillary units per 0.5 mL ($\log(4.0 \times 106) = 6.6$) with $n = 72$ observations, whereas regimen 4.3 has also $n = 72$ observations. Both data sets are listed in Table 6.13; see Bjerkedal (1960) and Kundu et al. (2008).

Human mortality

The Mexican Institute of Social Security studied the mortality of retired people on disability because it enables the calculation of long- and short-term financial estimation, such as the assessment of the reserve required to pay the minimum pensions. The data correspond to age at death (in years) of $n = 280$ retired women with temporary disabilities, which are incorporated in the Mexican public insurance system and who died during 2004. These data are given in Table 6.14 with frequency in parentheses and no parentheses when the frequency is one; see Balakrishnan et al. (2009a) and Sanhueza et al. (2009).

Insulating fluid

These data represent $n = 19$ observations of lifetimes (in minutes) to breakdown of an insulating fluid under 32 kV of voltage stress. The data set is displayed in Table 6.15; see Soofi et al. (1995) and Mills (1997).

Table 6.13 Survival times of guinea pigs

Regimen 4.3											
10	100	116	153	197	254	33	102	120	159	202	254
44	105	121	160	213	278	56	107	122	163	215	293
59	107	122	163	216	327	72	108	124	168	222	342
74	108	130	171	230	347	77	108	134	172	231	361
92	109	136	176	240	402	93	112	139	183	245	432
96	113	144	195	251	458	100	115	146	196	253	555
Regimen 6.6											
12	15	22	24	24	32	32	33	34	38	38	43
44	48	52	53	54	54	55	56	57	58	58	59
60	60	60	60	61	62	63	65	65	67	68	70
70	72	73	75	76	76	81	83	84	85	87	91
95	96	98	99	109	110	121	127	129	131	143	146
146	175	175	211	233	258	258	263	297	341	341	376

Lung cancer

Male patients with advanced inoperable lung cancer were randomized to either standard or test chemotherapy. It is of interest to explain the survival time (T, in days) of $n = 137$ patients with lung cancer (9 of them were censored, with censoring indicator $\delta = 1$: censoring; $\delta = 0$: noncensoring) by the covariables: (X_1) a measure, at randomization, of the patient's performance status (Karnofsky rating), where 10–30 means completely hospitalized, 40–60 means partially confinement, and 70–90 means to be able to take care of themselves; (X_2) time in months from diagnosis to randomization; (X_3) age in years; (X_4) prior therapy, a dichotomous variable taking the value 10 for yes and 0 (zero) for no; (X_5, X_6, X_7) histological type of tumor, which has the categories squamous, small cell, adeno, and large cell, making necessary the use of dummy variables given by $X_5 = 1, X_6 = 1$ and $X_7 = 1$, if the type of cancer cell is squamous, small, and adeno, respectively, and 0 otherwise; and (X_8) type of treatment, which takes the value 0 for standard chemotherapy and 1 for test chemotherapy. The data set is displayed in Table 6.16; see Kalbfleisch and Prentice (2002).

Multiple myeloma

It is of interest to relate the survival time (T, in months) to prognosic variables. Here, $n = 65$ patients with multiple myeloma were considered and treated with alkylating agents, of which 48 died during the study and 17 lived and therefore were censored (with censoring indicator $\delta = 1$:

Table 6.14 Age at death (in years) of Mexican retired women with temporary disabilities

22	24	25 (2)	27	28	29 (4)	30	31 (6)	32 (7)	33 (3)	34 (6)	35 (4)	36 (11)	37 (5)	38 (3)
39 (6)	40 (14)	41 (12)	42 (6)	43 (5)	44 (7)	45 (10)	46 (6)	47 (5)	48 (11)	49 (8)	50 (8)	51 (8)	52 (14)	53 (10)
54 (13)	55 (11)	56 (10)	57 (15)	58 (11)	59 (9)	60 (7)	61 (2)	62	63	64 (4)	65 (2)	66 (3)	71	74
75	79	86												

Table 6.15 Lifetimes (in minutes) of insulating fluids

0.19	0.78	0.96	1.31	2.78	3.16	4.15	4.67	4.85	6.50
7.35	8.01	8.27	12.06	31.75	32.52	33.91	36.71	72.89	

censoring; $\delta = 0$: noncensoring). The five prognosis variables are: (X_1) logarithm of a blood urea nitrogen measurement at diagnosis; (X_2) hemoglobin measurement at diagnosis; (X_3) age at diagnosis; (X_4) sex, 0 for male and 1 for female; and (X_5) serum calcium measurement at diagnosis. The data set is displayed in Table 6.17; see Krall et al. (1975), Lawless (1982), and Jin et al. (2003).

Machine valves

This data set corresponds to failure times of machine valves of an industrial process grouped in $m = 30$ samples of $n = 5$ each, which is displayed in Table 6.18; see Leiva et al. (2011b).

Shelf life of food products

This data set corresponds to the shelf life (in days) of a food product; see Gacula and Kubala (1975) and Folks and Chhikara (1978). The data set is displayed in Table 6.19.

6.3 SOFTWARE

`bs` *package*

An `R` package named `bs` was developed by Leiva et al. (2006) for the Birnbaum–Saunders distribution and can be downloaded from https://cran.r-project.org/src/contrib/Archive/bs. The `bs` package contains basic probabilistic functions, reliability indicators, and random number generators from this distribution. To deal with the computation of the probability density, cumulative distribution and quantile functions, the commands `dbs()`, `pbs()`, and `qbs()` must be used, respectively. The following instructions illustrate these commands:

```
dbs(3, alpha = 0.5, beta = 1.0, log = FALSE)
[1] 0.02133878
pbs(1, alpha = 0.5, beta = 1.0, log = FALSE)
[1] 0.5
qbs(0.5, alpha = 0.5, beta = 2.5, log = FALSE)
[1] 2.5
```

To conduct a lifetime analysis, the failure rate, failure rate average, reliability function, and cumulative failure rate have been implemented by the commands `frbs()`, `frabs()`, `rfbs()`, and `crfbs()`, respectively.

Table 6.16 Survival times and covariables of patients with lung cancer

T	δ	X_1	X_2	X_3	X_4	X_5	X_6	X_7	X_8	T	δ	X_1	X_2	X_3	X_4	X_5	X_6	X_7	X_8
72	1	60	7	69	0	1	0	0	0	999	1	90	12	54	10	1	0	0	1
411	1	70	5	64	10	1	0	0	0	112	1	80	6	60	0	1	0	0	1
228	1	60	3	38	0	1	0	0	0	87	0	80	3	48	0	1	0	0	1
126	1	60	9	63	10	1	0	0	0	231	0	50	8	52	10	1	0	0	1
118	1	70	11	65	10	1	0	0	0	242	1	50	1	70	0	1	0	0	1
10	1	20	5	49	0	1	0	0	0	991	1	70	7	50	10	1	0	0	1
82	1	40	10	69	10	1	0	0	0	111	1	70	3	62	0	1	0	0	1
110	1	80	29	68	0	1	0	0	0	1	1	20	21	65	10	1	0	0	1
314	1	50	18	43	0	1	0	0	0	587	1	60	3	58	0	1	0	0	1
100	0	70	6	70	0	1	0	0	0	389	1	90	2	62	0	1	0	0	1
42	1	60	4	81	0	1	0	0	0	33	1	30	6	64	0	1	0	0	1
8	1	40	58	63	10	1	0	0	0	25	1	20	36	63	0	1	0	0	1
144	1	30	4	63	0	1	0	0	0	357	1	70	13	58	0	1	0	0	1
25	0	80	9	52	10	1	0	0	0	467	1	90	2	64	0	1	0	0	1
11	1	70	11	48	10	1	0	0	0	201	1	80	28	52	10	1	0	0	1
30	1	60	3	61	0	0	1	0	0	1	1	50	7	35	0	1	0	0	1
384	1	60	9	42	0	0	1	0	0	30	1	70	11	63	0	1	0	0	1
4	1	40	2	35	0	0	1	0	0	44	1	60	13	70	10	1	0	0	1
54	1	80	4	63	10	0	1	0	0	283	1	90	2	51	0	1	0	0	1
13	1	60	4	56	0	0	1	0	0	15	1	50	13	40	10	1	0	0	1

Continued

Table 6.16 Survival times and covariables of patients with lung cancer–cont'd

T	δ	X_1	X_2	X_3	X_4	X_5	X_6	X_7	X_8	T	δ	X_1	X_2	X_3	X_4	X_5	X_6	X_7	X_8
123	0	40	3	55	0	0	1	0	0	25	1	30	2	69	0	0	1	0	1
97	0	60	5	67	0	0	1	0	0	103	0	70	22	36	10	0	1	0	1
153	1	60	14	63	10	0	1	0	0	21	1	20	4	71	0	0	1	0	1
59	1	30	2	65	0	0	1	0	0	13	1	30	2	62	0	0	1	0	1
117	1	80	3	46	0	0	1	0	0	87	1	60	2	60	0	0	1	0	1
16	1	30	4	53	10	0	1	0	0	2	1	40	36	44	10	0	1	0	1
151	1	50	12	69	0	0	1	0	0	20	1	30	9	54	10	0	1	0	1
22	1	60	4	68	0	0	1	0	0	7	1	20	11	66	0	0	1	0	1
56	1	80	12	43	10	0	1	0	0	24	1	60	8	49	0	0	1	0	1
21	1	40	2	55	10	0	1	0	0	99	1	70	3	72	0	0	1	0	1
18	1	20	15	42	0	0	1	0	0	8	1	80	2	68	0	0	1	0	1
139	1	80	2	64	0	0	1	0	0	99	1	85	4	62	0	0	1	0	1
20	1	30	5	65	0	0	1	0	0	61	1	70	2	71	0	0	1	0	1
31	1	75	3	65	0	0	1	0	0	25	1	70	2	70	0	0	1	0	1
52	1	70	2	55	0	0	1	0	0	95	1	70	1	61	0	0	1	0	1
287	1	60	25	66	10	0	1	0	0	80	1	50	17	71	0	0	1	0	1
18	1	30	4	60	0	0	1	0	0	51	1	30	87	59	10	0	1	0	1
51	1	60	1	67	0	0	1	0	0	29	1	40	8	67	0	0	1	0	1
122	1	80	28	53	0	0	1	0	0	24	1	40	2	60	0	0	0	1	1
27	1	60	8	62	0	0	1	0	0	18	1	40	5	69	10	0	0	1	1

Table 6.16 Survival times and covariables of patients with lung cancer–cont'd

T	δ	X_1	X_2	X_3	X_4	X_5	X_6	X_7	X_8	T	δ	X_1	X_2	X_3	X_4	X_5	X_6	X_7	X_8
54	1	70	1	67	0	0	1	0	0	83	0	99	3	57	0	0	0	1	1
7	1	50	7	72	0	0	1	0	0	31	1	80	3	39	0	0	0	1	1
63	1	50	11	48	0	0	1	0	0	51	1	60	5	62	0	0	0	1	1
392	1	40	4	68	0	0	1	0	0	90	1	60	22	50	10	0	0	1	1
10	1	40	23	67	10	0	1	0	0	52	1	60	3	43	0	0	0	1	1
8	1	20	19	61	10	0	0	1	0	73	1	60	3	70	0	0	0	1	1
92	1	70	10	60	0	0	0	1	0	8	1	50	5	66	0	0	0	1	1
35	1	40	6	62	0	0	0	1	0	36	1	70	8	61	0	0	0	1	1
117	1	80	2	38	0	0	0	1	0	48	1	10	4	81	0	0	0	1	1
132	1	80	5	50	0	0	0	1	0	7	1	40	4	58	0	0	0	1	1
12	1	50	4	63	10	0	0	1	0	140	1	70	3	63	0	0	0	1	1
162	1	80	5	64	0	0	0	1	0	186	1	90	3	60	0	0	0	1	1
3	1	30	3	43	0	0	0	1	0	84	1	80	4	62	10	0	0	1	1
95	1	80	4	34	0	0	0	1	0	19	1	50	10	42	0	0	0	1	1
177	1	50	16	66	10	0	0	0	0	45	1	40	3	69	0	0	0	1	1
162	1	80	5	62	0	0	0	0	0	80	1	40	4	63	0	0	0	1	1
216	1	50	15	52	0	0	0	0	0	52	1	60	4	45	0	0	0	0	1
553	1	70	2	47	0	0	0	0	0	164	1	70	15	68	10	0	0	0	1
278	1	60	12	63	0	0	0	0	0	19	1	30	4	39	10	0	0	0	1
12	1	40	12	68	10	0	0	0	0	53	1	60	12	66	0	0	0	0	1
260	1	80	5	45	0	0	0	0	0	15	1	30	5	63	0	0	0	0	1

Continued

Table 6.16 Survival times and covariables of patients with lung cancer–cont'd

T	δ	X_1	X_2	X_3	X_4	X_5	X_6	X_7	X_8	T	δ	X_1	X_2	X_3	X_4	X_5	X_6	X_7	X_8
200	1	80	12	41	10	0	0	0	0	43	1	60	11	49	10	0	0	0	1
156	1	70	2	66	0	0	0	0	0	340	1	80	10	64	10	0	0	0	1
182	0	90	2	62	0	0	0	0	0	133	1	75	1	65	0	0	0	0	1
143	1	90	8	60	0	0	0	0	0	111	1	60	5	64	0	0	0	0	1
105	1	80	11	66	0	0	0	0	0	231	1	70	18	67	10	0	0	0	1
103	1	80	5	38	0	0	0	0	0	378	1	80	4	65	0	0	0	0	1
250	1	70	8	53	10	0	0	0	0	49	1	30	3	37	0	0	0	0	1
100	1	60	13	37	10	0	0	0	0										

Table 6.17 Survival times and covariables of patients with multiple myeloma

T	δ	X_1	X_2	X_3	X_4	X_5	T	δ	X_1	X_2	X_3	X_4	X_5
1	0	2.218	9.4	67	0	10	26	0	1.230	11.2	49	1	11
1	0	1.940	12.0	38	0	18	32	0	1.322	10.6	46	0	9
2	0	1.519	9.8	81	0	15	35	0	1.114	7.0	48	0	10
2	0	1.748	11.3	75	0	12	37	0	1.602	11.0	63	0	9
2	0	1.301	5.1	57	0	9	41	0	1.000	10.2	69	0	10
3	0	1.544	6.7	46	1	10	42	0	1.146	5.0	70	1	9
5	0	2.236	10.1	50	1	9	51	0	1.568	7.7	74	0	13
5	0	1.681	6.5	74	0	9	52	0	1.000	10.1	60	1	10
6	0	1.362	9.0	77	0	8	54	0	1.255	9.0	49	0	10
6	0	2.114	10.2	70	1	8	58	0	1.204	12.1	42	1	10
6	0	1.114	9.7	60	0	10	66	0	1.447	6.6	59	0	9
6	0	1.415	10.4	67	1	8	67	0	1.322	12.8	52	0	10
7	0	1.978	9.5	48	0	10	88	0	1.176	10.6	47	1	9
7	0	1.041	5.1	61	1	10	89	0	1.322	14.0	63	0	9
7	0	1.176	11.4	53	1	13	92	0	1.431	11.0	58	1	11
9	0	1.724	8.2	55	0	12	4	1	1.945	10.2	59	0	10
11	0	1.114	14.0	61	0	10	4	1	1.924	10.0	49	1	13
11	0	1.230	12.0	43	0	9	7	1	1.114	12.4	48	1	10
11	0	1.301	13.2	65	0	10	7	1	1.532	10.2	81	0	11
11	0	1.508	7.5	70	0	12	8	1	1.079	9.9	57	1	8
11	0	1.079	9.6	51	1	9	12	1	1.146	11.6	46	1	7
13	0	0.778	5.5	60	1	10	11	1	1.613	14.0	60	0	9
14	0	1.398	14.6	66	0	10	12	1	1.398	8.8	66	1	9
15	0	1.602	10.6	70	0	11	13	1	1.663	4.9	71	1	9
16	0	1.342	9.0	48	0	10	16	1	1.146	13.0	55	0	9
16	0	1.322	8.8	62	1	10	19	1	1.322	13.0	59	1	10
17	0	1.230	10.0	53	0	9	19	1	1.322	10.8	69	1	10
17	0	1.591	11.2	68	0	10	28	1	1.230	7.3	82	1	9
18	0	1.447	7.5	65	1	8	41	1	1.756	12.8	72	0	9
19	0	1.079	14.4	51	0	15	53	1	1.114	12.0	66	0	11
19	0	1.255	7.5	60	1	9	57	1	1.255	12.5	66	0	11
24	0	1.301	14.6	56	1	9	77	1	1.079	14.0	60	0	12
25	0	1.000	12.4	67	0	10							

Table 6.18 Failure times (in hours, t_i) of valves (i)

Index i	t_i					Index i	t_i				
1	4.55	4.99	3.62	3.52	3.77	16	2.53	4.16	3.78	3.77	1.72
2	1.93	3.95	4.10	4.16	1.61	17	3.41	3.10	6.02	1.09	2.92
3	2.22	1.73	5.10	4.52	4.06	18	2.85	4.46	3.17	2.50	3.91
4	2.71	2.45	4.60	2.09	1.90	19	3.16	3.70	2.61	2.65	3.42
5	2.91	5.68	4.33	3.51	3.24	20	2.54	4.77	1.63	2.64	3.59
6	2.20	5.66	3.71	3.35	1.61	21	3.61	2.13	5.08	2.01	1.92
7	2.82	5.22	3.75	3.50	3.31	22	3.16	4.20	2.32	2.44	1.62
8	2.76	4.40	3.13	1.55	3.70	23	2.96	6.09	3.78	2.29	4.16
9	4.98	4.05	4.00	7.20	3.18	24	2.47	3.49	3.38	4.45	2.61
10	4.88	2.71	3.51	3.15	4.81	25	3.55	3.35	3.18	4.75	8.72
11	4.50	1.95	3.41	2.87	1.90	26	1.35	2.50	2.51	4.20	3.50
12	3.07	4.02	4.17	4.33	4.06	27	2.30	2.26	2.22	1.60	9.70
13	2.39	2.91	3.09	3.15	2.52	28	3.71	3.06	1.53	2.45	6.40
14	2.92	4.25	3.02	2.26	5.72	29	9.48	1.72	4.20	3.37	5.58
15	2.56	4.38	1.24	2.62	1.92	30	1.90	2.56	4.28	3.18	1.94

Table 6.19 Shelf life (in days) of food products

24	24	26	26	32	32	33	33	33	35	41	42	43
47	48	48	48	50	52	54	55	57	57	57	57	61

To obtain random numbers, three generators were developed using Algorithms 5, 6, and 7 presented in Chapter 2 by the commands `rbs1()`, `rbs2()`, and `rbs3()`, respectively. Also, a common command, named `rbs()`, for obtaining random numbers from the more suitable generator, depending on the desired setting, has been developed. The command `rbs()` automatically selects the most appropriate method. For more details about effectiveness, efficiency, and most suitable selection of these three generators, the interested reader is referred to Leiva et al. (2008b). The following instructions illustrate these commands:

```
rbs3(n = 6, alpha = 0.5, beta = 1.0)
[1] 0.7372910 0.4480005 1.8632176
[4] 0.9728011 1.2675537 0.2252379

rbs(n = 6, alpha = 0.5, beta = 1.0)
[1] 0.5905414 1.1378133 1.1664306
[4] 1.3187935 1.2609212 1.8212990
```

Another group of commands related to estimation, graphical analysis and goodness-of-fit for the Birnbaum–Saunders distribution are also available in the bs package. To estimate the shape (α) and scale (β) parameters of this distribution, the three estimation methods (maximum likelihood, least squares, and modified moments) described in Chapter 3 have been implemented by the commands `est1bs()`, `est2bs()`, and `est3bs()`, respectively. Next, an example related to the use of these commands is presented. A sample from a Birnbaum–Saunders distribution with data rounded to two decimal numbers and saved as a slot of the R-class as `sample` is generated by using the following instruction with its result indicated below:

```
sample <- round(rbs(10, 0.5, 1), 2)
[1] 0.77 0.94 0.85 1.61 0.88 0.94 0.83 0.63 1.79 3.05
```

The estimation of parameters from each method is carried out with the following instructions and their results are provided below each of them:

```
est1bs(sample)
$beta.start
[1] 1.106449
$alpha
[1] 0.4706615
$beta
[1] 1.107785
$converge
[1] "TRUE"
$iteration
[1] 3

est2bs(sample)
$alpha
[1] 0.4659694
$beta
[1] 1.156344
$r.squared
[1] 0.6071716

est3bs(sample)
$alpha
[1] 0.4706597
$beta
[1] 1.106449
```

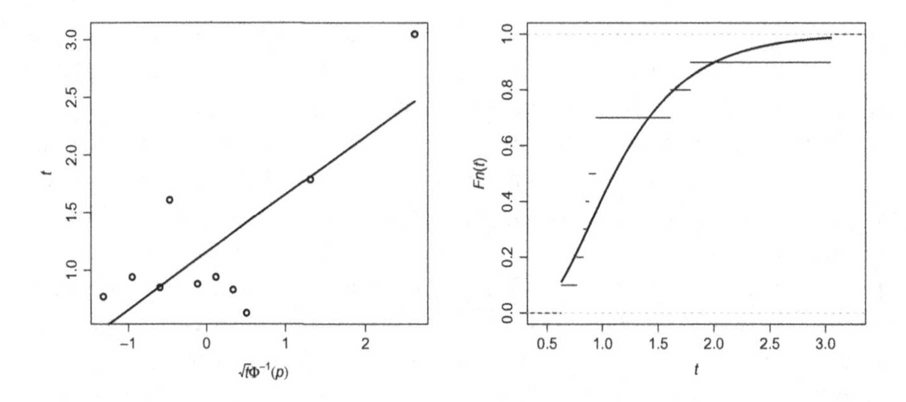

Figure 6.1 Plots of probability obtained by regression for the Birnbaum–Saunders distribution (left) and of the empirical and theoretical distribution functions (right) for simulated data.

If the instruction

```
estimates <- est1bs(sample)
```

is considered, the estimates of α and β can be saved as follows:

```
alpha <- estimate$alpha
beta  <- estimate$beta
```

Then, estimates of α and β can be recovered. If the command `est2bs(sample)` is replaced by `gmbs(sample)`, in addition to the estimates of α and β obtained from the graphical method introduced in Chapter 2, `gmbs(sample)` provides the corresponding probability plot displayed in Figure 6.1(left). This figure shows a good fit of the Birnbaum–Saunders distribution to the simulated data. The Kolmogorov–Smirnov goodness-of-fit test for the Birnbaum–Saunders distribution is available in the bs package by means of the instruction:

```
ksbs(sample, alternative=c("two.sided"), plot=TRUE)
One-sample Kolmogorov-Smirnov test
data: sample
D = 0.33658, p-value = 0.2073
alternative hypothesis: two-sided
```

Note that the argument `plot = TRUE` in the above instruction provides Figure 6.1(right), which shows the coherence between the empirical and theoretical cumulative distribution functions.

Also, by using the invariance property of the maximum likelihood estimators, we can estimate the mean, variance, and coefficients of variation, skewness, and kurtosis of the Birnbaum–Saunders distribution by the following instruction with its corresponding results given below:

```
indicatorsbs(sample)
Alpha =  0.4706615
Beta  =  1.107785
Mean      =  1.368481
Variance =  0.347125
Coefficient of variation =  0.478814
Coefficient of skewness  =  1.375995
Coefficient of kurtosis  =  6.087569
```

Next, some other implemented graphical commands are described. Firstly, the instructions:

```
x <- seq(0,4,by=0.01)
y <- dbs(x,alpha=0.3,beta=1.0)
z <- dbs(x,alpha=0.5,beta=1.0)
w <- dbs(x,alpha=1,beta=1.0)
h <- dbs(x,alpha=2,beta=1.0)
plot(c(0,4),c(0,1.5),type="n",xlab="f(x)",ylab="x")
lines(x,y,type="l",lty=1,lwd=4,add="T",col=1)
lines(x,z,type="l",lty=2,lwd=4,add="T",col=1)
lines(x,w,type="l",lty=3,lwd=4,add="T",col=1)
lines(x,h,type="l",lty=1,lwd=4,add="T",col=8)
legend("topright",lty=c(1,2,3,1),
       lwd=c(4,4,4,4),col=c(1,1,1,8),
       c(expression(alpha*"=0.3"),
         expression(alpha*"=0.5"),
         expression(alpha*"=1.0"),
         expression(alpha*"=2.0")))
```

produce Figure 2.1 of Chapter 2 related to the Birnbaum–Saunders density, whereas Figure 2.2 of Chapter 2 corresponding to its failure rate is produced modifying the instruction

```
h <- dbs(x,alpha=2,beta=1.0)
```

by the instruction

```
h <- frbs(x, alpha = 2, beta = 1.0)
```

The bs package has also implemented commands related to the model selection criteria AIC, BIC, and HQIC for the Birnbaum–Saunders distribution. By using data sample, these commands and their results are:

```
sicbs(sample)
[1] 0.9548382
aicbs(sample)
[1] 0.9245797
hqbs(sample)
[1] 0.8913862
```

Values of the AIC, BIC and HQIC allow us to compare the Birnbaum–Saunders distribution to other distributions. Also, goodness-of-fit methods based on moments, such as β_1–β_2 and δ_2–δ_3 charts, can be obtained using the bs package for the Birnbaum–Saunders distribution; see Chapter 2. As mentioned, these charts of fit are particularly useful when various data sets are collected. For example, in environmental sciences, one frequently finds random variables measured per hour at different sampling points. Then, for each year, one has a monthly data set available at each sampling point. Thus, it is possible to compute estimates of β_1 (the square of the coefficient of skewness), β_2 (the coefficient of kurtosis), δ_2 (the square of the coefficient of variation), and δ_3 (the coefficient of skewness), for each month at each sampling point. In this way, the pairs $(\hat{\beta}_1, \hat{\beta}_2)$ and $(\hat{\delta}_2, \hat{\delta}_3)$ are plotted inside of the theoretical β_1–β_2 and δ_2–δ_3 charts, respectively. The commands `fitbetabs()` and `fitdeltabs()` have been developed to add these charts to the scatter-plot of the pairs $(\hat{\beta}_1, \hat{\beta}_2)$ and $(\hat{\delta}_2, \hat{\delta}_3)$, respectively. To illustrate this methodology, 12 samples have been simulated using the instruction

```
rbs(n = 30, alpha, beta = 1.0)
```

for α from 0.2 to 2.4 by 0.2, which can represent, for example, daily data for each month during one year. These results have been saved in the matrix `samplei`, for $i = 1, \ldots, 12$. The sample values of δ_2, δ_3, β_1, and β_2 are computed and saved in the vectors x and y. The simulated data sets have been fitted to the mentioned charts. Thus, plots of Figure 6.2 (left and right, respectively) are obtained by using the instructions:

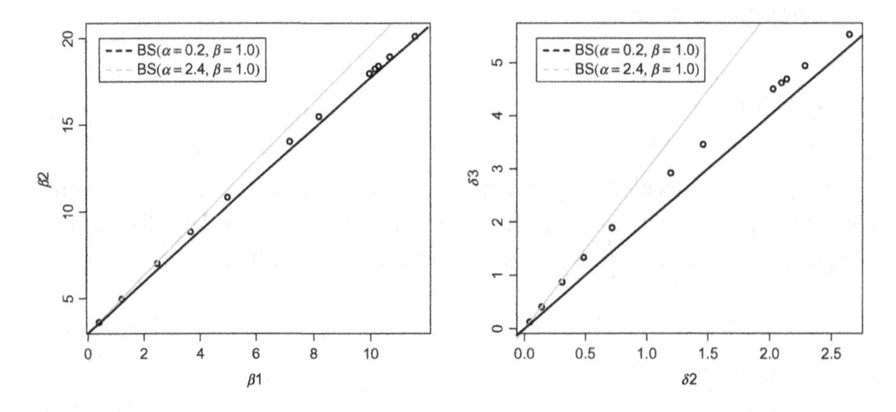

Figure 6.2 *β_1–β_2 (left) and δ_2–δ_3 (right) charts for the indicated distributions and 12 simulated samples with their sample values for $\beta_1, \beta_2, \delta_2$ and δ_3 shown as circles.*

```
rbs(n = 30, alpha, beta = 1.0)
plot(x, y, xlab = expression(beta*1),
           ylab = expression(beta*2), col = 4)
fitbetabs(0.2, 2)
fitbetabs(2.4, 3)

plot(x, y, xlab = expression(delta*2),
           ylab = expression(delta*3), col = 4)
fitdeltabs(0.2, 2)
fitdeltabs(2.4, 3)
```

gbs *package*

Several of the data sets presented in Section 6.2 are incorporated in the bs package. An R package named gbs was developed by Barros et al. (2009) for Birnbaum–Saunders distributions (including a generalization of this) and can be downloaded from https://cran.r-project.org/src/contrib/Archive/gbs. Next, we discuss commands of this package related to the Birnbaum–Saunders distribution, but not to its generalization. Similarly to the bs package, the gbs package has implemented basic probabilistic functions, a random number generator, the maximum likelihood estimation method for uncensored and censored data, as well as goodness-of-fit and diagnostic techniques for the Birnbaum–Saunders distribution. To deal with the computation of the probability density, cumulative distribution and quantile functions, the instructions

```
dgbs(x, kernel = "normal")
pgbs(x, kernel = "normal")
qgbs(x, kernel = "normal")
```

must be used, respectively. In any case, the instruction `kernel = "normal"` is assumed by default in the gbs package. For this package, only those commands distinct from the bs package are discussed next. A differentiating aspect of the gbs package is related to exploratory data analysis, which can be used for any distribution.

As an example to illustrate the differentiating aspects of the gbs package in relation to the bs package, the real-world fatigue-life data `psi31` given in Table 6.1 and implemented in the gbs package are considered. Specifically, in order to carry out a descriptive summary, the command `descriptiveSummary()` was implemented. The results for data `psi31` are presented in Table 6.20, including the standard deviation (SD) and coefficients of variation (CV), skewness (CS), and kurtosis (CK). These results are obtained by the instruction:

```
descriptiveSummary(psi31)
```

Table 6.20 Descriptive statistics for psi31

n	Minimum	Median	Mean	Maximum	SD	CV	CS	CK	Range
101	70	133.00	133.73	212.00	22.36	0.17	0.33	3.97	142.00

Also, this command returns the mode of the data. However, in this example, this value was not considered. The instruction

```
histgbs(psi31, mainTitle = "", densityLine = TRUE,
        xLabel = "Lifetime with stress of 31.000 psi",
        colourHistogram = 0, colourBoxPlot = 0)
```

simultaneously produces a box-plot and a histogram for data `psi31`. The box-plot may be suppressed by the instruction `boxPlot = FALSE`. Figure 6.4(left) shows these graphs. From Table 6.20 and Figure 6.4(left), note that a positively skewed distribution with moderate kurtosis and some atypical values are detected. Thus, these fatigue-life data can be well modeled by the Birnbaum–Saunders distribution. By using the invariance property of the maximum likelihood estimators, the estimated Birnbaum–Saunders probability density function may be plotted on the histogram adding in `histgbs()` the argument `densityLine = TRUE`; see Figure 6.4(left) obtained with data `psi31`, from which is noted that the Birnbaum–Saunders distribution shows an excellent fit to these data.

Another differentiating aspect of the gbs package in relation to the bs package is the estimation of parameters with censored data, as well as their inference and diagnostics. The command `acigbs()` computes an $[1 - \eta] \times 100\%$ approximate confidence region for the parameters α and β of the Birnbaum–Saunders distribution as defined in Chapter 3. In addition, it also provides graphs of these regions, which are reported by approximate simultaneous confidence intervals. By using data `psi31`, these intervals and the approximate confidence region for the parameters α and β of the Birnbaum–Saunders distribution are computed by means of the following instructions, with their corresponding results below, and shown in Figure 6.3(left):

```
acigbs(psi31, kernel="normal", confLevel=95,
       chart=c(0.1, 0.2, 125, 140), colourRegion=1,
       colourEstimates=1)
$alphaEstimate
[1] 0.1703847
$alphaAci
[1] 0.14 0.20
```

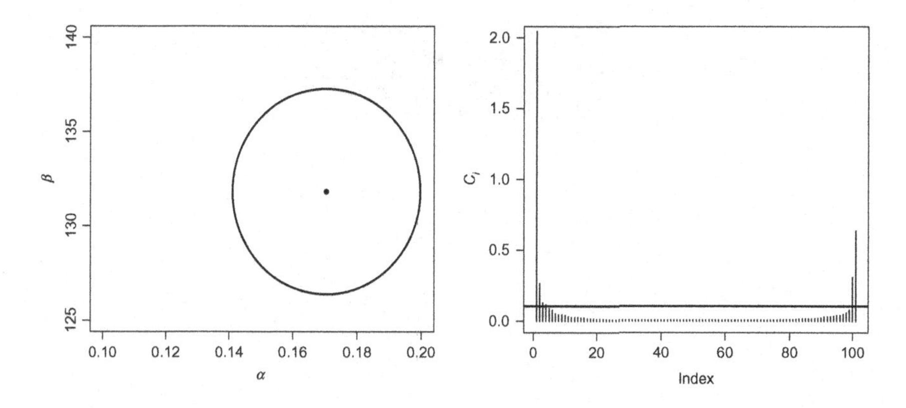

Figure 6.3 Confidence region for α and β (left) and local influence index plot (right) with `psi31` data.

```
$betaEstimate
[1] 131.8188
$betaAci
[1] 126.37 137.27
```

Diagnostics based on the total influence method can be carried out numerically and graphically by the command `diagnosticsgbs()`. Thus, by using data `psi31`, the instruction

```
diagnosticsgbs(psi31, yRange = c(0,2))
```

produces the graph presented in Figure 6.3(right), which indicates a potential influence on the maximum likelihood estimates of the cases: #1, #2, #100, and #101 of `psi31`. In order to establish the magnitude of the changes produced on the maximum likelihood estimates of α and β, when the potentially influential cases are removed, the command `rcgbs()` was implemented. This command computes the relative changes in percentage of each estimated parameter and is defined as $\mathrm{RC}_{\theta_j} = |(\hat{\theta}_j - \hat{\theta}_{j(\mathrm{I})})/\hat{\theta}_j| \times 100\%$, where $\hat{\theta}_{j(\mathrm{I})}$ denotes the maximum likelihood estimate of θ_j after the set I of cases has been removed. Table 6.21 shows these relative changes obtained by using the instruction:

```
rcgbs(psi31, casesRemoved=1)
```

Analogously, the instructions `casesRemoved = 2` and `casesRemoved = c(1, 2, 100, 101)` allow us to remove the case #2 and cases $\{\#1, \#2, \#100, \#101\}$, respectively. Observe that more noticeable changes are detected for $\hat{\alpha}$ instead of $\hat{\beta}$.

Table 6.21 Relative change (in %) for the indicated parameter and removed case

Removed case(s)	$\hat{\alpha}$	$\hat{\beta}$
#1	6.95	0.67
#2	2.09	0.39
#100	2.30	0.40
#101	3.57	0.49
{#1, #2, #100, #101}	16.13	0.17

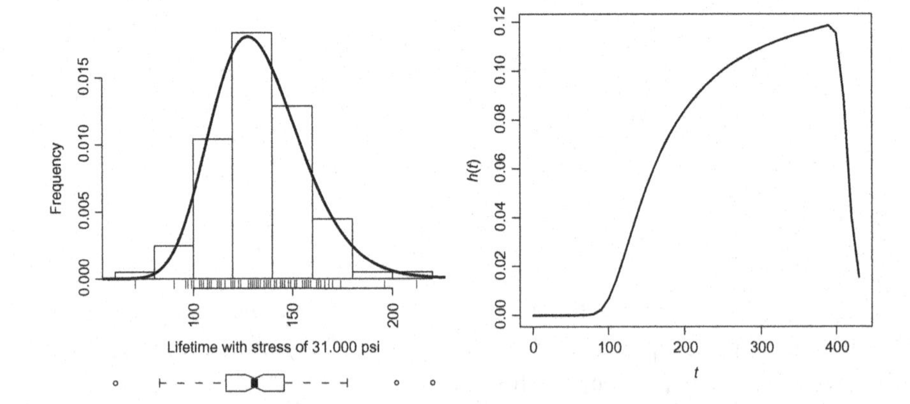

Figure 6.4 Histogram and box-plot with estimated Birnbaum–Saunders density (left) and estimated Birnbaum–Saunders failure rate (right) for `psi31` data.

In order to carry out a reliability analysis, the gbs package has implemented the commands `rfgbs()` and `frgbs()` to compute the reliability function and failure rate, respectively. Once again by using the invariance property of the maximum likelihood estimators, the failure rate of the Birnbaum–Saunders distribution can be estimated and plotted by means of the instructions:

```
t  <- seq(0.1, 440, by=10)
ht <- frgbs(t, alpha=alpha, beta=beta)
plot(t, ht, type="l", main="",
     lty=c(1,2,1), col=cbind(1,1,8), lwd=2.0,
     xlab="t", ylab="h(t)", cex.main=2)
```

Figure 6.4(right) shows the estimated Birnbaum–Saunders failure rate. In addition, an estimate of the change point of the corresponding failure rate has been implemented in a new version of the gbs package (which is available under request to the author) by the command `ecpfrgbs()`. Thus, the estimated change point of the Birnbaum–Saunders failure rate can be obtained by the instruction:

```
ecpfrgbs(psi31)
$changePoint
[1] 391.7115
$maximum
[1] 0.1185348
```

Note that the estimated change point coincides with that observed in Figure 6.4(right), as well as the maximum value of the estimated Birnbaum–Saunders failure rate.

Censored data

Some progress has been reached with censored data and goodness of fit in relation to the `bs` and `gbs` packages. The maximum likelihood estimates of the parameters α and β based on right type-II censored data can be obtained by using the command `mlebsc()`. Thus, for data `psi31` censored at approximately 20%, that is, with a number of failures $r = 80$ and $m = 21$ censored data, the estimated parameters are $\hat{\alpha} = 0.1751$ and $\hat{\beta} = 132.2525$; see Barros et al. (2014). Based on Algorithm 6 for goodness of fit with censored data presented in Chapter 5, the observed values of the statistics are $\text{ad} = 0.6075$, $\text{cm} = 0.1106$, and $\text{ks} = 0.9166$. For a level of significance $1 - \varrho = 0.05$, the critical values for the corresponding test statistics are $\text{ad}_{0.95} = 0.7313$, $\text{cm}_{0.95} = 0.1224$, and $\text{ks}_{0.95} = 0.8860$. Therefore, the hypothesis indicating that the data come from the Birnbaum–Saunders distribution at 5% is not rejected, when the AD* and CM* statistics are used. However, this decision changes when the KS* statistic is considered, which is expected because the corresponding test based on this statistic has less power; see Barros et al. (2014). Thus, decisions taken from AD* or CM* statistics must be considered. When Algorithm 7 is used, the observed values $\text{ad}_{r,n} = 1.482$ and $\text{cm}_{r,n} = 0.3076$ are obtained, with critical values $\text{ad}_{0.95} = 2.243$ and $\text{cm}_{0.95} = 0.4251$, respectively, implying H_0 is not rejected at a 5% significance level.

For uncensored data `psi31`, acceptance regions at a 5% significance level to test goodness of fit in the Birnbaum–Saunders distribution can be plotted by using Algorithm 8. Figure 6.5 shows PP plots with the p-value of the Kolmogorov–Smirnov test, which allows us to support the assumption of a Birnbaum–Saunders distribution for these data.

6.4 MODELING AND DIAGNOSTICS WITH REAL-WORLD DATA

Modeling and diagnostics with uncensored data

(Example with data of Table 6.5)

As discussed by Rieck and Nedelman (1991) and supported by the exploratory analysis presented in Figure 6.6 for data of Table 6.5, we see

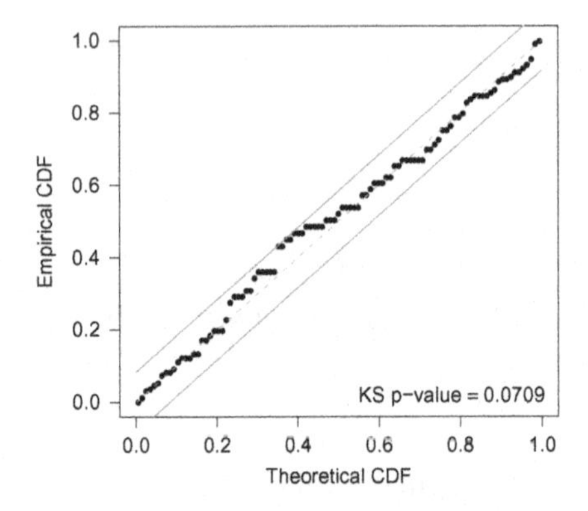

Figure 6.5 PP plot with acceptance bands at a 5% level for the Birnbaum–Saunders distribution using `psi31` *data.*

that an exponential regression model is adequate for these data, that is, the log-linear model

$$Y_i = \log(T_i) = \eta_0 + \eta_1 \log(x_i) + \varepsilon_i, \quad i = 1, 2, \ldots, 46, \tag{6.1}$$

where $\varepsilon_i \sim \text{log-BS}(\alpha, 0)$. The maximum likelihood estimates of $\boldsymbol{\theta} = (\alpha, \boldsymbol{\eta}^\top)^\top = (\alpha, \eta_0, \eta_1)^\top$ are (with estimated standard errors in parenthesis): $\hat{\eta}_0 = 12.280\,(0.403)$, $\hat{\eta}_1 = -1.671\,(0.112)$, and $\hat{\alpha} = 0.410\,(-)$. Note both regression coefficients are significant at 5%.

Next, an analysis of local influence for the data set given in Table 6.5 is carried out based on the Birnbaum–Saunders regression model given in Equation (6.1) and $\ell(\boldsymbol{\theta}|\boldsymbol{\omega})$ given in Equations (4.31), (4.33), and (4.36), according to the indicated perturbation scheme. In each scheme, the analysis is made for the vector $\boldsymbol{\theta}$ and, then, for the partitions α and $\boldsymbol{\eta}$ of the parameter vector $\boldsymbol{\theta}$, that is, $\boldsymbol{\theta} = (\theta_1, \boldsymbol{\theta}_2^\top)^\top$, where $\theta_1 = \alpha$ and $\boldsymbol{\theta}_2 = \boldsymbol{\eta} = (\eta_0, \eta_1)^\top$, in this example.

Case-weight perturbation

Based on Equation (6.1) and with $\ell(\boldsymbol{\theta}|\boldsymbol{\omega})$ given as in Equation (4.31), it follows that $C_{l_{\max}}(\boldsymbol{\theta}) = 1.312$. For α and $\boldsymbol{\eta}$, the largest eigenvalues are given by $C_{l_{\max}}(\alpha) = 0.593$ and $C_{l_{\max}}(\boldsymbol{\eta}) = 1.288$. Figure 6.7 presents the eigenvector corresponding to $\boldsymbol{\eta}$.

Thus, from Figure 6.7, the cases #4 and #5 exercise an important influence on $\hat{\boldsymbol{\eta}}$. We notice a less pronouncing influence of cases #2, #3, #12, and #46. We remove the most influential cases (#2, #3, #4, #5, #12, #46) and refit the model. The relative change is defined by

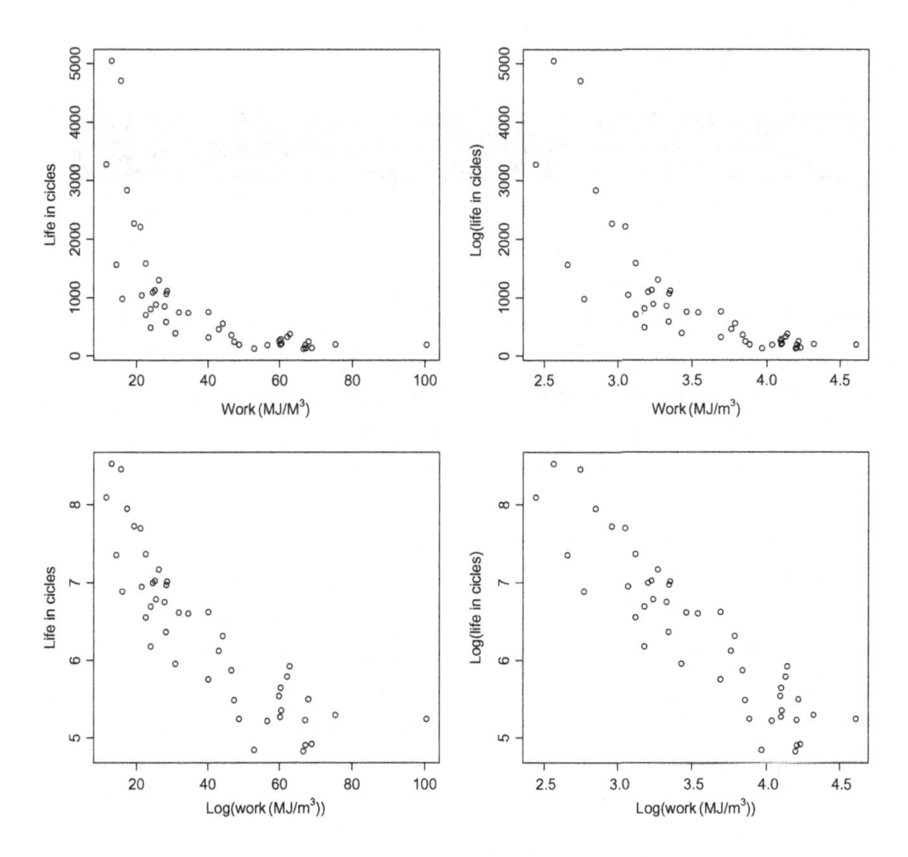

Figure 6.6 Scatter-plots of X versus T (first row, left), X versus log(*T*) *(first row, right),* log(*X*) *versus T (second row, left), and* log(*X*) *versus* log(*T*) *(second row, right) for biaxial fatigue data.*

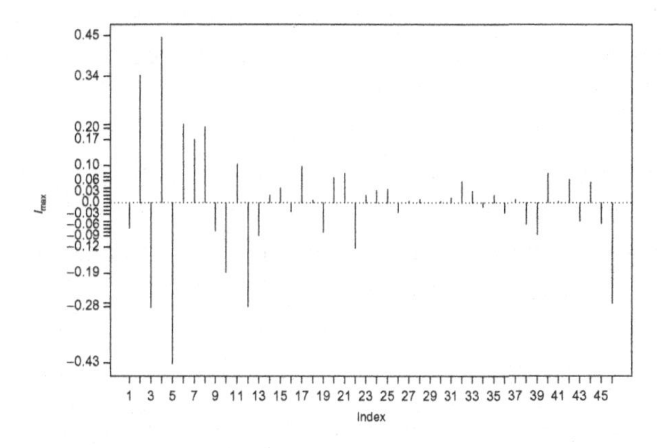

Figure 6.7 Local influence index plots for the estimated model coefficients under case perturbation in the Birnbaum–Saunders model fitted to biaxial fatigue data. From Galea et al. (2004), with permission.

$$\mathrm{RC}_{\eta_j} = \left| \frac{\hat{\eta}_j - \hat{\eta}_{j(i)}}{\hat{\eta}_j} \right|, \quad j = 0, 1, \ i = 1, \ldots, 46, \tag{6.2}$$

Table 6.22 Relative changes on $\hat{\eta}$ after dropping the indicated case for the Birnbaum–Saunders model fitted to the biaxial fatigue data set

Dropped case	$\hat{\eta}_0$	Relative change	p-Value	$\hat{\eta}_1$	Relative change	p-Value
None	12.280	0.000	<0.001	−1.671	0.000	<0.001
2	12.123	0.013	<0.001	−1.630	0.025	<0.001
3	12.406	0.010	<0.001	−1.703	0.019	<0.001
4	12.085	0.016	<0.001	−1.620	0.031	<0.001
5	12.468	0.015	<0.001	−1.719	0.029	<0.001
12	12.379	0.008	<0.001	−1.694	0.014	<0.001
46	12.471	0.016	<0.001	1.729	0.035	<0.001

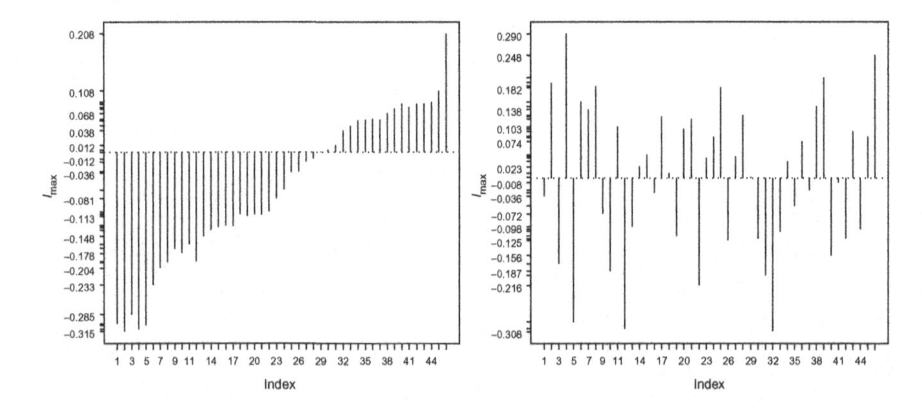

Figure 6.8 Local influence index plot for the estimated model coefficients (left) and the estimated shape parameter (right) under response perturbation in the Birnbaum–Saunders model fitted to biaxial fatigue data. From Galea et al. (2004), with permission.

where $\hat{\eta}_{j(i)}$ denotes the estimate of η_j after case i is removed. Relative changes on the estimates of η_0 and η_1 are given in Table 6.22. As can be noticed, cases #4 and #46 change more than 3% the estimate of η_1, but the significance of η_1 is not modified.

Response perturbation

Similarly, based on the model given in Equation (6.1) and $\ell(\boldsymbol{\theta}|\boldsymbol{\omega})$ given in Equation (4.33), it follows that $C_{l_{\max}}(\boldsymbol{\theta}) = 13.014$. The largest eigenvalues are given by $C_{l_{\max}}(\alpha) = 13.009$ and $C_{l_{\max}}(\boldsymbol{\eta}) = 6.451$. Figure 6.8(left) indicates that cases with large and small values for X exercise larger influence on $\hat{\boldsymbol{\eta}}$ when $\boldsymbol{y}$ is perturbed. This is expected because such cases are typically of high leverage. Figure 6.8(right) does not present cases with high influence, whose behavior is similar to that from $\hat{\boldsymbol{\theta}}$.

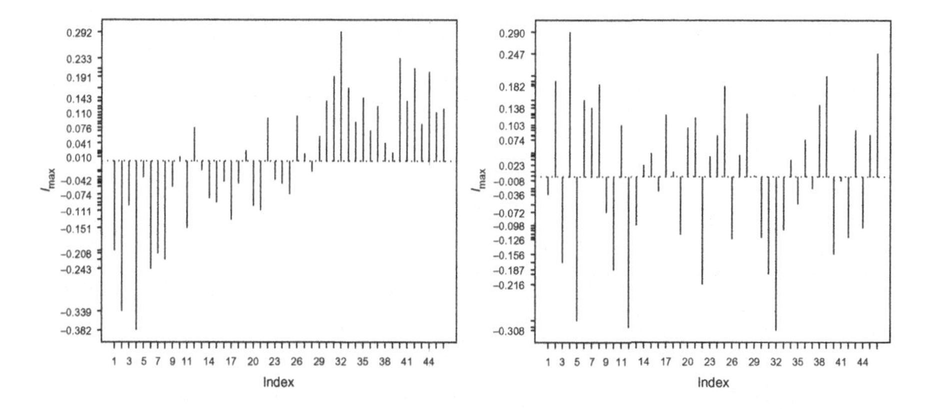

Figure 6.9 Local influence index plot for the estimated model coefficients (left) and the estimated shape parameter (right) under covariable perturbation in the Birnbaum–Saunders model fitted to biaxial fatigue data. From Galea et al. (2004), with permission.

Explanatory variable perturbation

Based on the model given in Equation (6.1) and $\ell(\boldsymbol{\theta}|\boldsymbol{\omega})$ in Equation (4.36), it follows that $C_{l_{\max}}(\boldsymbol{\theta}) = 126.062$. Analogously, the largest eigenvalues are given by $C_{l_{\max}}(\alpha) = 36.348$ and $C_{l_{\max}}(\boldsymbol{\eta}) = 111.094$. The interpretation of Figure 6.9, under perturbation scheme of the covariable, is similar to that of Figure 6.8, under the perturbation scheme of the response.

Now, an analysis of total local influence is presented, according to the perturbation scheme indicated. We use the data described in Table 6.5, the model given in Equation (6.1) and the expression for C_i provided in Equation (4.28), with $i = 1, \ldots, 46$. In each scheme, the analysis is made for the vector $\boldsymbol{\theta}$ and, then, for the partitions α and $\boldsymbol{\eta}$ of this vector.

Case-weight perturbation

Similarly to the graphical plots of local influence shown in Figures 6.7–6.9, cases #2, #3, #4, #5, #12, and #46 appear as the most influential. However, based on Figure 6.10, the total local influence method detects the case #32 as influential, but the local influence method not. It is important to highlight that the total local influence method seems to detect more cases than the local influence method.

Response perturbation

Figure 6.11 exhibits the index plot of C_i under the response perturbation scheme. Here, the same cases are detected, but Figure 6.11(left) presents a tendency indicating that cases with smaller and larger values for the covariable X appear with a high influence. A similar tendency is detected in the local influence index plots.

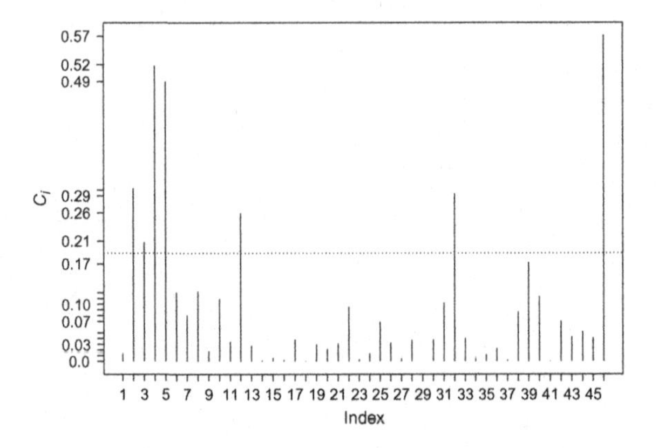

Figure 6.10 Local influence index plot for the estimated model coefficients under case perturbation in the Birnbaum–Saunders model fitted to biaxial fatigue data. From Galea et al. (2004), with permission.

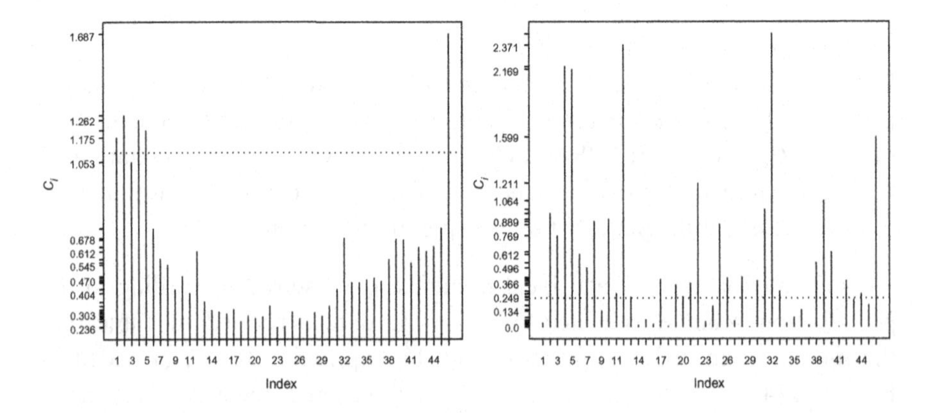

Figure 6.11 Total local influence index plots for the estimated model coefficients (left) and estimated shape parameter (right) under response perturbation in the Birnbaum–Saunders model fitted to biaxial fatigue data. From Galea et al. (2004), with permission.

Explanatory variable perturbation

The index plot of C_i under local perturbation in the covariable X is similar to the tendencies observed in Figures 6.10 and 6.11.

Generalized leverage analysis

Figure 6.12 shows the index plot of $\hat{\text{GL}}_{ii}$ using the model given in Equation (6.1). The generalized leverage graph presented in Figure 6.12 confirms the tendencies observed under local and total influence methods when the response is perturbed. Observations with large and small values for X tend to have a high influence on their own fitted values. Note that an outstanding influence is detected for the case #46. The $\hat{\text{GL}}_{ii}$ graph is very similar to that given in Figure 6.11(left).

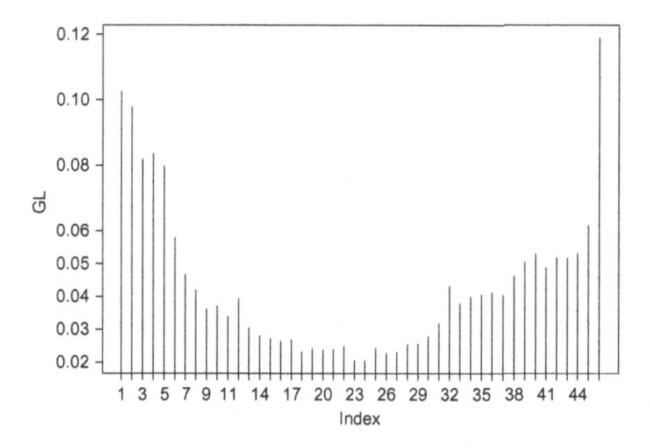

Figure 6.12 Generalized leverage plot for the estimated model coefficients in the Birnbaum–Saunders model fitted to biaxial fatigue data. From Galea et al. (2004), with permission.

Table 6.23 Maximum likelihood estimate and its corresponding standard error for the indicated parameter in the Birnbaum–Saunders model fitted to myeloma data

Parameter	η_0	η_1	η_2	η_3	η_4	η_5	α
Estimate	4.500	−1.596	0.142	0.010	0.209	−0.141	1.082
Estimated standard error	1.282	0.435	0.050	0.013	0.284	−0.069	0.112

Modeling and diagnostics with censored data
(Example with data of Table 6.17)
Consider the model

$$Y_i = \eta_0 + \eta_1 x_{i1} + \eta_2 x_{i2} + \eta_3 x_{i3} + \eta_4 x_{i4} + \eta_5 x_{i5} + \varepsilon_i, \quad i = 1, \ldots, 65, \quad (6.3)$$

where $\varepsilon_i \sim$ log-BS$(\alpha, 0)$ is the model error and Y_i the logarithm of the lifetime of the patient i. The maximum likelihood estimate of

$$\boldsymbol{\theta} = (\alpha, \boldsymbol{\eta}^\top)^\top = (\alpha, \eta_0, \eta_1, \eta_2, \eta_3, \eta_4, \eta_5)^\top$$

and the corresponding estimated standard errors are presented in Table 6.23. From this table, note that covariables X_3 and X_4 are not significant at a 10% level.

To detect possible departures from the assumption of log-Birnbaum–Saunders errors in the model given in Equation (6.3), Figure 6.13 presents the normal probability plots for deviance component and martingale residuals with generated envelopes; for more details about the envolope plots, the interested reader is referred to Atkinson (1981) and Leiva et al. (2007).

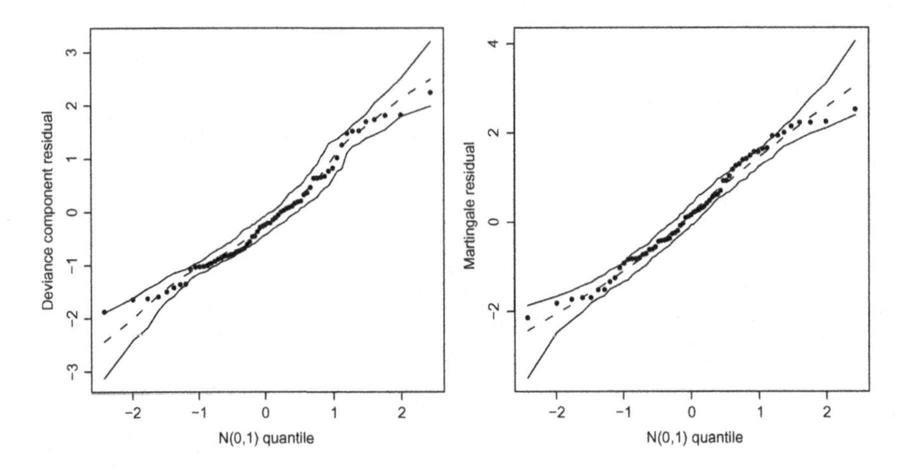

Figure 6.13 Normal probability plots for the deviance component (left) and martingale (right) residuals with envelopes in the Birnbaum–Saunders model fitted to myeloma data.

Note that the graphs in Figure 6.13 do not present unusual features so that this assumption does not seem to be unsuitable.

In order to detect atypical and high leverage cases, the index of $\hat{\mathrm{GL}}_{ii}$ is plotted in Figure 6.14. In this figure, no case appears as a possible atypical case. In Figure 6.14, case #2 appears with a high leverage point. This case corresponds to the youngest patient, 38 years old, male and just 1 month of survival since the treatment. His logarithm of blood urea nitrogen and hemoglobin measurements are greater than the average of the 65 patients, whereas his serum calcium measurement is very high, the largest among the patients.

Figures 6.15–6.19 show the index plots of $|\boldsymbol{l}_{\max}|$ for $\boldsymbol{\theta}$, α, and $\boldsymbol{\eta}$, where $\boldsymbol{\theta} = (\theta_1, \boldsymbol{\theta}_2^\top)^\top$, with $\theta_1 = \alpha$ and $\boldsymbol{\theta}_2 = \boldsymbol{\eta} = (\eta_0, \eta_1, \ldots, \eta_5)^\top$, under the scheme indicated. As noted from these figures, case #40 appears as the most influential in all the graphs. This case corresponds to a male patient, 74 years old and survival time since the treatment of 51 months. His logarithm of blood urea nitrogen and serum calcium measurements are greater than the average of the 65 patients, whereas his hemoglobin measurement is below the average. Other cases appear with some outstanding influence on the parameter estimates. For example, cases #3, #5, #44, and #48 are potentially influential on $\hat{\alpha}$ and $\hat{\boldsymbol{\eta}}$ under the case perturbation scheme. When only the censored cases are perturbed, the cases #62, #64, and #65 are detected as potentially influential; see Figure 6.18. Some index plots of $|\boldsymbol{l}_{\max}|$ for

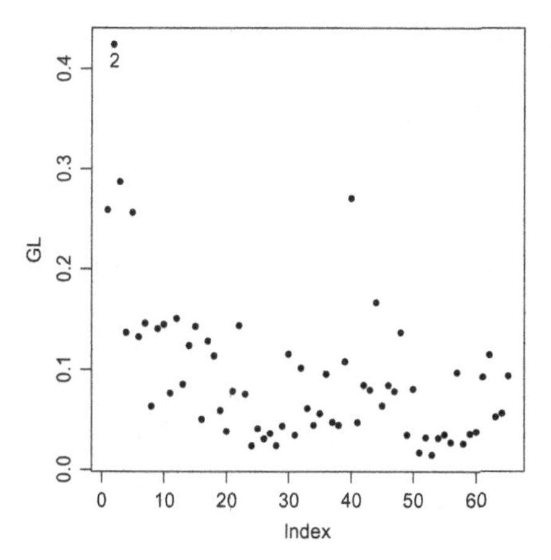

Figure 6.14 Generalized leverage plot for the Birnbaum–Saunders model fitted to myeloma data.

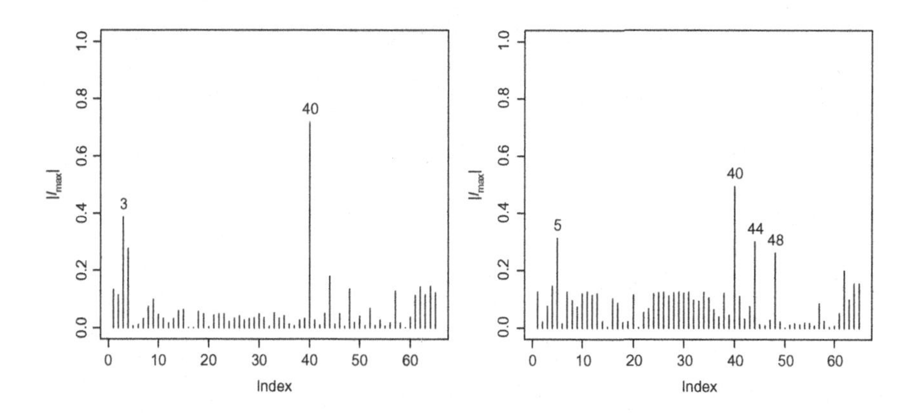

Figure 6.15 Local influence index plots for the estimated model coefficients (left) and the estimated shape parameter (right) under the case perturbation in the Birnbaum–Saunders model fitted to myeloma data.

η and/or α have been omitted due to their similarity with those given in Figures 6.15–6.19.

The index plots of C_i for $\boldsymbol{\theta}$, α, and η under each perturbation scheme (omitted here) confirmed the outstanding influence of cases #3, #5, #40, #44, #48, #62, #64, and #65, which also were detected in the previous index plots of $|\boldsymbol{l}_{\max}|$. Therefore, our diagnostic analysis detects as potentially influential the cases #2, #3, #5, #40, #44, #48, #62, #64, and #65. To reveal the impact of these nine cases on the parameter estimates, we refit the model under the following situations. First, individually each one of these nine cases

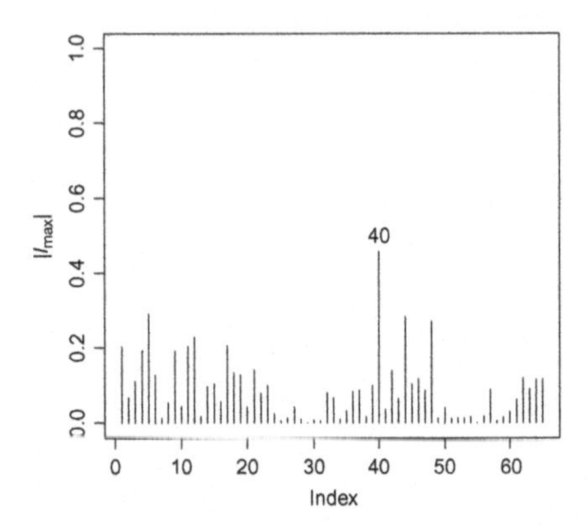

Figure 6.16 Local influence index plots for the estimated model coefficients (a) and the estimated shape parameter (b) under the response perturbation in the Birnbaum–Saunders model fitted to myeloma data.

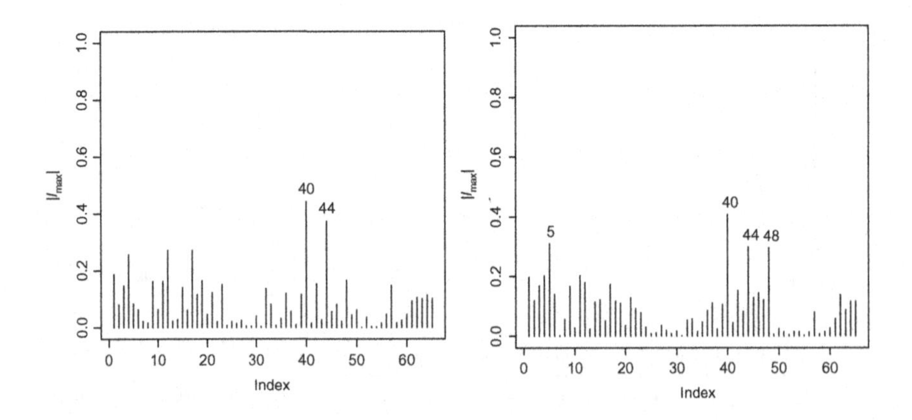

Figure 6.17 Local influence index plots for the estimated model coefficients (left) and the estimated shape parameter (right) under the covariable perturbation in the Birnbaum–Saunders model fitted to myeloma data.

is removed. Then, once dropping from the "set A" (original data set) the total of potentially influential cases, the model is refitted (named "set B"). Table 6.24 presents the estimates of the model parameters for the sets A and B. In Table 6.25, the relative changes of each parameter estimate are presented, which are defined in an analogous way as to what was presented in Equation (6.2) replacing the index i by a "set I" of indexes. Thus,

$$\mathrm{RC}_{\theta_j} = \left| \frac{\hat{\theta}_j - \hat{\theta}_{j(\mathrm{I})}}{\hat{\theta}_j} \right| \times 100\%, \quad j = 1, \ldots, p+1,$$

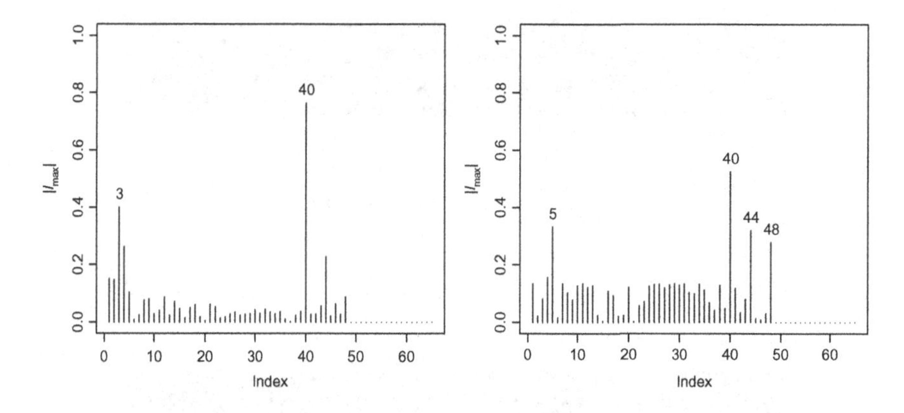

Figure 6.18 Local influence index plots for the estimated model coefficients (left) and the estimated shape parameter (right) under noncensored perturbation in the Birnbaum–Saunders model fitted to myeloma data.

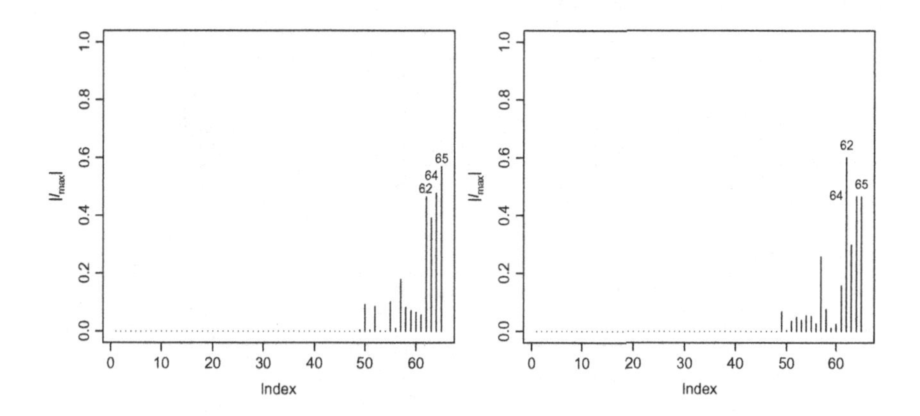

Figure 6.19 Local influence index plots for the estimated model coefficients (left) and the estimated shape parameter (right) under censored perturbation in the Birnbaum–Saunders model fitted to myeloma data.

where $\hat{\theta}_{j(I)}$ denotes the maximum likelihood estimate of θ_j after the "set I" of cases has been removed. In addition, Table 6.25 shows the p-values of the corresponding t-test.

As noted from Table 6.24, inferences do not change at a 10% significance level when the set A or the set B is considered. This indicates that the covariables X_3 and X_4 should be removed from the model. However, looking at Table 6.25, note that the elimination of cases #2 and #3 makes the

Table 6.24 Maximum likelihood estimate and p-value of the corresponding t-test for the indicated parameter in the Birnbaum–Saunders model fitted to myeloma data

Set	Parameter	η_0	η_1	η_2	η_3	η_4	η_5
A	Estimate	4.500	−1.596	0.142	0.010	0.209	−0.141
	p-value	<0.001	<0.001	0.004	0.455	0.461	0.042
B	Estimate	5.685	−1.617	0.099	−0.005	0.119	−0.142
	p-value	<0.001	<0.001	0.037	0.692	0.634	0.105

Table 6.25 Relative changes (in %) and p-value of the corresponding t-test for the Birnbaum–Saunders model fitted to myeloma data

Set I	$\hat{\eta}_0$	$\hat{\eta}_1$	$\hat{\eta}_2$	$\hat{\eta}_3$	$\hat{\eta}_4$	$\hat{\eta}_5$	$\hat{\alpha}$
A−{case 2}	4	6	1	54	3	27	0
p-value	0.001	0.001	0.005	0.753	0.473	0.201	–
A−{case 3}	15	−3	3	−62	−4	30	1
p-value	0.005	<0.001	0.005	0.250	0.438	0.188	–
A−{case 5}	−20	−3	25	42	68	−9	4
p-value	<0.001	<0.001	0.046	0.660	0.814	0.023	–
A−{case 40}	−14	0	−14	71	−12	−34	6
p-value	<0.001	<0.001	<0.001	0.826	0.377	0.011	–
A−{case 44}	12	−1	−14	−22	−46	12	3
p-value	0.002	<0.001	0.001	0.350	0.283	0.063	–
A−{case 48}	−5	−2	5	4	52	−7	3
p-value	<0.001	<0.001	0.006	0.460	0.723	0.029	–
A−{case 62}	−5	−6	9	23	−7	11	2
p-value	<0.001	<0.001	0.009	0.562	0.422	0.065	–
A−{case 64}	−2	3	5	14	−9	−1	1
p-value	<0.001	<0.001	0.006	0.513	0.415	0.037	–
A−{case 65}	−2	6	8	5	−9	−6	1
p-value	<0.001	<0.001	0.008	0.471	0.414	0.031	–
B	-26	−1	30	148	43	−1	24
p-value	<0.001	<0.001	0.037	0.692	0.634	0.105	–

covariable X_5 to be nonsignificant. The significance of this variable was masked by these cases. Therefore, it should also be removed from the model. Thus, the final model becomes given by

$$y_i = \eta_0 + \eta_1 x_{i1} + \eta_2 x_{i2} + \varepsilon_i, \quad i = 1, \ldots, 65. \tag{6.4}$$

The maximum likelihood estimates of the parameters of the model established in (6.4) (with estimated standard errors in parenthesis) are: $\hat{\eta}_0 = 4.409$ (0.772), $\hat{\eta}_1 = -1.869$ (0.423), $\hat{\eta}_2 = 0.109$ (0.049), and $\hat{\alpha} = 1.140$ (0.118).

We may interpret the estimated coefficients of the final model as follows. The expected survival time should approximately decrease

$$85\% \, ((1 - \exp(-1.869)) \times 100\%)$$

as the logarithm of blood urea nitrogen measurement increases one unit of this covariable, keeping the hemoglobin measurement fixed. Similarly, the expected survival time should approximately increase 12% ($\exp(0.109) \times 100\%$) as the hemoglobin measurement increases one unit of this covariable, keeping the logarithm of blood urea nitrogen measurement fixed.

BIBLIOGRAPHY

Aarset, M., 1987. How to identify a bathtub hazard rate. IEEE Trans. Reliab. 36, 106–108.

Abramowitz, M., Stegun, I., 1972. Handbook of Mathematical Functions. Dover Press, New York.

Achcar, J., 1993. Inference for the Birnbaum-Saunders fatigue life model using Bayesian methods. Comput. Stat. Data Anal. 15, 367–380.

Achcar, J., Espinosa, M., 1992. Use of Bayesian methods in accelerated life tests considering a log-linear model for the Birnbaum–Saunders distribution. Rev. Bras. Estat. 52, 47–68.

Ahmad, L., 1988. Jackknife estimation for a family of life distributions. J. Stat. Comput. Simul. 29, 211–223.

Ahmed, S., Castro-Kuriss, C., Leiva, V., Flores, E., Sanhueza, A., 2010. Truncated version of the Birnbaum–Saunders distribution with an application in financial risk. Pak. J. Stat. 26, 293–311.

Ang, A., Tang, W., 1975. Probability Concepts in Engineering Planning and Design, vol. I. Wiley, New York.

ASTM International, 2013. Standard Terminology Relating to Fatigue and Fracture Testing. ASTM-E1823-13. ASTM International, West Conshohocken, US. http://www.astm.org/Standards/E1823.

Athayde, E., Azevedo, C., Leiva, V., Sanhueza, A., 2012. About Birnbaum–Saunders distributions based on the Johnson system. Commun. Stat. Theory Methods 41, 2061–2079.

Atkinson, A. C., 1981. Two graphical display for outlying and influential observations in regression. Biometrika 68, 13–20.

Azevedo, C., Leiva, V., Athayde, E., Balakrishnan, N., 2012. Shape and change point analyses of the Birnbaum–Saunders-t hazard rate and associated estimation. Comput. Stat. Data Anal. 56, 3887–3897.

Balakrishnan, N., Leiva, V., López, J., 2007. Acceptance sampling plans from truncated life tests based on the generalized Birnbaum–Saunders distribution. Commun. Stat. Simul. Comput. 36, 643–656.

Balakrishnan, N., Leiva, V., Sanhueza, A., Cabrera, E., 2009a. Mixture inverse Gaussian distribution and its transformations, moments and applications. Statistics 43, 91–104.

Balakrishnan, N., Leiva, V., Sanhueza, A., Vilca, F., 2009b. Estimation in the Birnbaum–Saunders distribution based on scale-mixture of normals and the EM-algorithm. Stat. Oper. Res. Trans. 33, 171–192.

Balakrishnan, N., Gupta, R., Kundu, D., Leiva, V., Sanhueza, A., 2011. On some mixture models based on the Birnbaum–Saunders distribution and associated inference. J. Stat. Plan. Inference 141, 2175–2190.

Barlow, R., Proschan, F., 1965. Mathematical Theory of Reliability. Wiley, New York.

Barlow, W., Prentice, R., 1988. Residuals for relative risk regression. Biometrika 75, 65–74.

Barros, M., Paula, G., Leiva, V., 2008. A new class of survival regression models with heavy-tailed errors: robustness and diagnostics. Lifetime Data Anal. 14, 316–332.

Barros, M., Paula, G., Leiva, V., 2009. An R implementation for generalized Birnbaum–Saunders distributions. Comput. Stat. Data Anal. 53, 1511–1528.

Barros, M., Leiva, V., Ospina, R., Tsuyuguchi, A., 2014. Goodness-of-fit tests for the Birnbaum–Saunders distribution with censored reliability data. IEEE Trans. Reliab. 63, 543–554.

Bartlett, M., Kendall, D., 1946. The statistical analysis of variance-heterogeneity and the logarithmic transformation. J. R. Stat. Soc. Ser. B 8, 128–138.

Belsley, D., Kuh, E., Welsch, R., 1980. Regressions Diagnostics: Identifying Influential Data and Sources Collinearity. Wiley, New York.

Bhattacharyya, G., Fries, A., 1982. Fatigue failure models: Birnbaum–Saunders versus inverse Gaussian. IEEE Trans. Device Mater. Reliab. 31, 439–440.

Bhatti, C., 2010. The Birnbaum–Saunders autoregressive conditional duration model. Math. Comput. Simul. 80, 2062–2078.

Birnbaum, Z., Saunders, S., 1958. A statistical model for life-length of materials. J. Am. Stat. Assoc. 53, 151–160.

Birnbaum, Z., Saunders, S., 1968. A probabilistic interpretation of Miner's rule. SIAM J. Appl. Math. 16, 637–652.

Birnbaum, Z., Saunders, S., 1969a. A new family of life distributions. J. Appl. Probab. 6, 319–327.

Birnbaum, Z., Saunders, S., 1969b. Estimation for a family of life distributions with applications to fatigue. J. Appl. Probab. 6, 328–347.

Birnbaum, Z., Esary, J., Marshall, A., 1966. Stochastic characterization of wear-out for components and systems. Ann. Math. Stat. 37, 816–825.

Bjerkedal, T., 1960. Acquisition of resistance in guinea pigs infected with different doses of virulent tubercle bacilli. Am. J. Hyg. 72, 130–148.

Bourguignon, M., Silva, R., Cordeiro, G., 2014. A new class of fatigue life distributions. J. Stat. Comput. Simul. 84, 2619–2635.

Bourguignon, M., Leão, J., Leiva, V., Santos-Neto, M., 2015. The transmuted Birnbaum–Saunders distribution. Working Paper. Available at http://victorleiva.cl/archivos/leiva_art/article_BLLS.pdf.

Brown, M., Miller, K., 1978. Biaxial Fatigue Data. University of Sheffield, Sheffield, UK.

Castillo, N., Gomez, H., Bolfarine, H., 2011. Epsilon Birnbaum–Saunders distribution family: properties and inference. Stat. Papers 52, 871–883.

Castro-Kuriss, C., 2011. On a goodness-of-fit test for censored data from a location-scale distribution with application. Chilean J. Stat. 37, 115–136.

Castro-Kuriss, C., Kelmansky, D., Leiva, V., Martínez, E., 2009. A new goodness-of-fit test for censored data with an application in monitoring processes. Commun. Stat. Simul. Comput. 38, 1161–1177.

Castro-Kuriss, C., Kelmansky, D., Leiva, V., Martínez, E., 2010. On a goodness-of-fit test for normality with unknown parameters and type-II censored data. J. Appl. Stat. 37, 1193–1211.

Castro-Kuriss, C., Leiva, V., Athayde, E., 2014. Graphical tools to assess goodness-of-fit in non-location-scale distributions. Colombian J. Stat. 37, 341–365.

Chang, D., Tang, L., 1993. Reliability bounds and critical time for the Birnbaum–Saunders distribution. IEEE Trans. Device Mater. Reliab. 42, 464–469.

Chang, D., Tang, L., 1994a. Graphical analysis for Birbaum-Saunders distribution. Microelectron. Reliab. 34, 17–22.

Chang, D., Tang, L., 1994b. Percentile bounds and tolerance limits of the Birnbaum–Saunders distribution. Commun. Stat. Theory Methods 23, 2853–2863.

Chatterjee, S., Hadi, A., 1988. Sensitivity Analysis in Linear Regression. Wiley, New York.

Chen, G., Balakrishnan, N., 1995. A general purpose approximate goodness-of-fit test. J. Qual. Technol. 27, 154–161.

Chhikara, R., Folks, J., 1977. The inverse Gaussian distribution as a lifetime model. Technometrics 19, 461–468.

Chhikara, R., Folks, J., 1989. The Inverse Gaussian Distribution—Theory, Methodology, and Applications. Marcel Dekker, New York.

Collins, J., 1981. Failure of Materials in Mechanical Design. Wiley, New York.

Cook, D., 1986. Assessment of local influence. J. R. Stat. Soc. Ser. B 48, 133–169.

Cook, D., Weisberg, S., 1982. Residuals and Influence in Regression. Chapman and Hall, London, UK.

Cordeiro, M., Lemonte, J., 2011. The β-Birnbaum–Saunders distribution: an improved distribution for fatigue life modeling. Comput. Stat. Data Anal. 55, 1445–1461.

Cordeiro, M., Lemonte, J., 2014. The exponentiated generalized Birnbaum–Saunders distribution. Appl. Math. Comput. 247, 762–779.

Cox, D., Hinkley, D., 1974. Theoretical Statistics. Chapman and Hall, London, UK.

Cox, D., Oakes, D., 1984. Analysis of Survival Data. Chapman and Hall, London, UK.

Cramér, H., 1946. Mathematical Methods of Statistics. Princeton University Press, Princeton.

Cramér, H., Leadbetter, M., 1967. Stationary and Related Stochastic Processes. Wiley, New York.

Cysneiros, A., Cribari-Neto, F., Araujo, C., 2008. On Birnbaum–Saunders inference. Comput. Stat. Data Anal. 52, 4939–4950.

D'Agostino, C., Stephens, M., 1986. Goodness-of-Fit Techniques. Marcel Dekker, New York.

Davis, D., 1952. An analysis of some failure data. J. Am. Stat. Assoc. 47, 113–150.

Davison, A., Gigli, A., 1989. Deviance residuals and normal scores plots. Biometrika 76, 211–221.

Desmond, A., 1985. Stochastic models of failure in random environments. Canadian J. Stat. 13, 171–183.

Desmond, A., 1986. On the relationship between two fatigue life models. IEEE Trans. Reliab. 35, 167–169.

Díaz-García, J., Leiva, V., 2003. Doubly non-central t and F distributions obtained under singular and non-singular elliptic distributions. Commun. Stat. Theory Methods 32, 11–32.

Díaz-Garcia, J., Leiva, V., 2005. A new family of life distributions based on elliptically contoured distributions. J. Stat. Plan. Inference 128, 445–457.

Díaz-Garcia, J., Galea, M., Leiva, V., 2003. Influence diagnostics for elliptical multivariate linear regression models. Commun. Stat. Theory Methods 32, 625–641.

Dupuis, D., Mills, J., 1998. Robust estimation of the Birnbaum–Saunders distribution. IEEE Trans. Reliab. 1, 88–95.

Efron, B., Hinkley, D., 1978. Assessing the accuracy of the maximum likelihood estimator: observed vs. expected Fisher information. Biometrika 65, 457–487.

Elderton, W., 1927. Frequency Curves and Correlations. Harren, Washington.

Engelhardt, M., Bain, L., Wright, F., 1981. Inferences on the parameters of the Birnbaum–Saunders fatigue life distribution based on maximum likelihood estimation. Technometrics 23, 251–256.

Epstein, B., Sobel, M., 1953. Life testing. J. Am. Stat. Assoc. 48, 486–502.

Epstein, B., Sobel, M., 1954a. Some theorems relevant to life testing from an exponential distribution. Ann. Math. Stat. 25, 373–381.

Epstein, B., Sobel, M., 1954b. Truncated life tests in the exponential case. Ann. Math. Stat. 25, 555–564.

Epstein, B., Sobel, M., 1955. Sequential life tests in the exponential case. Ann. Math. Stat. 26, 82–93.

Esary, J., Marshall, A.W., Proschan, F., 1973. Shock models and wear processes. Ann. Probab. 1, 627–649.

Escobar, L., Meeker, W., 1992. Assesing local influence in regression analysis with censored data. Biometrics 48, 507–528.

Ferreira, M., Gomes, M., Leiva, V., 2012. On an extreme value version of the Birnbaum–Saunders distribution. Revstat-Stat. J. 10, 181–210.

Fierro, R., Leiva, V., Ruggeri, F., Sanhueza, A., 2013. On a Birnbaum–Saunders distribution arising from a non-homogeneous Poisson process. Stat. Probab. Lett. 83, 1233–1239.

Fischer, T., Kamps, U., 2011. On the existence of transformations preserving the structure of order statistics in lower dimensions. J. Stat. Plan. Inference 141, 536–548.

Folks, J., Chhikara, R., 1978. The inverse Gaussian distribution and its statistical application. J. R. Stat. Soc. Ser. B 20, 263–289.

Fox, E., Gavish, B., Semple, J., 2008. A general approximation to the distribution of count data with applications to inventory modeling. Working Paper available at http://dx.doi.org/10.2139/ssrn.979826.

Fréchet, M., 1927. Sur la loi de probabilité de l'écart maximum. Ann. Soc. Polon. Math. 6, 93–116.

Freudental, A., Gumbel, E., 1954. Minimun life in fatigue. J. Am. Stat. Assoc. 49, 575–597.

Freudenthal, A., Shinozuka, M., 1961. Structural Safety Under Conditions of Ultimate Load Failure and Fatigue. Wright Air Development Division, Wright Air Force Base, Ohio.

From, S., Li, L., 2006. Estimation of the parameters of the Birnbaum–Saunders distribution. Commun. Stat. Theory Methods 35, 2157–2169.

Gacula, M., Kubala, J., 1975. Statistical models for shelf life failures. J. Food Sci. 40, 404–409.

Galea, M., Paula, G., Bolfarine, H., 1997. Local influence in elliptical linear regression models. Statistician 46, 71–79.

Galea, M., Riquelme, M., Paula, G., 2000. Diagnostics methods in elliptical linear regression models. Braz. J. Probab. Stat. 14, 167–184.

Galea, M., Leiva, V., Paula, G., 2004. Influence diagnostics in log-Birnbaum–Saunders regression models. J. Appl. Stat. 31, 1049–1064.

Gomes, M.I., Ferreira, M., Leiva, V., 2012. The extreme value Birnbaum–Saunders model and its moments and application in biometry. Biom. Lett. 49, 81–94.

Gómez, H., Olivares-Pacheco, J., Bolfarine, H., 2009. An extension of the generalized Birnbaum–Saunders distribution. Stat. Probab. Lett. 79, 331–338.

Gradshteyn, I., Randzhik, I., 2000. Table of Integrals, Series, and Products. Academic Press, New York.

Guiraud, P., Leiva, V., Fierro, R., 2009. A non-central version of the Birnbaum–Saunders distribution for reliability analysis. IEEE Trans. Reliab. 58, 152–160.

Gumbel, E., 1958. Statistics of Extremes. Columbia University Press, New York.

Ho, J.W., 2012. Parameter estimation for the Birnbaum–Saunders distribution under an accelerated degradation test. Eur. J. Ind. Eng. 6, 644–665

Holmen, J., 1979. Fatigue of concrete by constant and variable amplitude loading. Division of Concrete Structures, The Norwegian Institute of Technology, University of Trondheim, Trondheim, Norway. http://dx.doi.org/10.14359/6402.

Hsu, Y., Pearn, W., Wu, P., 2008. Capability adjustment for gamma processes with mean shift consideration in implementing Six Sigma program. Eur. J. Oper. Res. 119, 517–529.

Jin, X., Kawczak, J., 2003. Birnbaum–Saunders and lognormal kernel estimators for modelling durations in high frequency financial data. Ann. Econ. Finance 4, 103–124.

Jin, Z., Lin, D., Wei, L., Ying, Z., 2003. Rank-based inference for the accelerated failure time model. Biometrika 90, 341–353.

Johnson, N., Kotz, S., Kemp, A., 1993. Univariate Discrete Distributions. Wiley, New York.

Johnson, N., Kotz, S., Balakrishnan, N., 1995. Continuous Univariate Distributions, vol. 2. Wiley, New York.

Kalbfleish, J., Prentice, R., 1980. The Statistical Analysis of Failure Time Data. Wiley, New York.

Kalbfleisch, J., Prentice, R., 2002. The Statistical Analysis of Failure Time Data. Wiley, New York.

Kao, J., 1959. A graphical estimation of mixed Weibull parameters in life-testing of electron tubes. Technometrics 1, 389–407.

Kass, R., Raftery, A., 1995. Bayes factors. J. Am. Stat. Assoc. 90, 773–795.

Kendall, M., Stuart, A., 1974. Advanced Theory of Statistics, vol. 2/3. MacMillan, New York.

Klein, J., Moeschberger, M., 1997. Survival Analysis: Techniques for Censored and Truncated Data. Springer, New York.

Kotz, S., Leiva, V., Sanhueza, A., 2010. Two new mixture models related to the inverse Gaussian distribution. Methodol. Comput. Appl. Probab. 12, 199–212.

Krall, J., Uthoff, V., Harley, J., 1975. A step-up procedure for selecting variables associated with survival. Biometrics 31, 49–57.

Kundu, D., Kannan, N., Balakrishnan, N., 2008. On the hazard function of Birnbaum–Saunders distribution and associated inference. Comput. Stat. Data Anal. 52, 2692–2702.

Lange, K., 2001. Numerical Analysis for Statisticians. Springer, New York.

Lawless, J., 1982. Statistical Models and Methods for Lifetime Data. Wiley, New York.

Lawrance, A., 1988. Regression transformation diagnostics using local influence. J. Am. Stat. Assoc. 84, 125–141.

Leiva, V., Hernandez, H., Riquelme, M., 2006. An new package for the Birnbaum–Saunders distribution. R J. 6, 35–40.

Leiva, V., Barros, M., Paula, G., Galea, M., 2007. Influence diagnostics in log-Birnbaum–Saunders regression models with censored data. Comput. Stat. Data Anal. 51, 5694–5707.

Leiva, V., Barros, M., Paula, G., Sanhueza, A., 2008a. Generalized Birnbaum–Saunders distribution applied to air pollutant concentration. Environmetrics 19, 235–249.

Leiva, V., Hernandez, H, Sanhueza, A., 2008b. An R package for a general class of inverse Gaussian distributions. J. Stat. Softw. 26, 1–21.

Leiva, V., Riquelme, M., Balakrishnan, N., Sanhueza, A., 2008c. Lifetime analysis based on the generalized Birnbaum–Saunders distribution. Comput. Stat. Data Anal. 52, 2079–2097.

Leiva, V., Sanhueza, A., Sen, P., Paula, G., 2008d. Random number generators for the generalized Birnbaum–Saunders distribution. J. Stat. Comput. Simul. 78, 1105–1118.

Leiva, V., Sanhueza, A., Silva, A., Galea, M., 2008e. A new three-parameter extension of the inverse Gaussian distribution. Stat. Probab. Lett. 78, 1266–1273.

Leiva, V., Sanhueza, A., Angulo, J.M., 2009. A length-biased version of the Birnbaum–Saunders distribution with application in water quality. Stoch. Environ. Res. Risk Assess. 23, 299–307.

Leiva, V., Sanhueza, A., Kotz, S., Araneda, N., 2010a. A unified mixture model based on the inverse Gaussian distribution. Pak. J. Stat. 26, 445–460.

Leiva, V., Vilca, F., Balakrishnan, N., Sanhueza, A., 2010b. A skewed sinh-normal distribution and its properties and application to air pollution. Commun. Stat. Theory Methods 39, 426–443.

Leiva, V., Athayde, E., Azevedo, C., Marchant, C., 2011a. Modeling wind energy flux by a Birnbaum–Saunders distribution with unknown shift parameter. J. Appl. Stat. 38, 2819–2838.

Leiva, V., Soto, G., Cabrera, E., Cabrera, G., 2011b. New control charts based on the Birnbaum–Saunders distribution and their implementation. Colombian J. Stat. 34, 147–176.

Leiva, V., Ponce, M., Marchant, C., Bustos, O., 2012. Fatigue statistical distributions useful for modeling diameter and mortality of trees. Colombian J. Stat. 35, 349–367.

Leiva, V., Marchant, C., Saulo, H., Aslam, M., Rojas, F., 2014a. Capability indices for Birnbaum–Saunders processes applied to electronic and food industries. J. Appl. Stat. 41, 1881–1902.

Leiva, V., Rojas, E., Galea, M., Sanhueza, A., 2014b. Diagnostics in Birnbaum–Saunders accelerated life models with an application to fatigue data. Appl. Stoch. Model. Bus. Ind. 30, 115–131.

Leiva, V., Santos-Neto, M., Cysneiros, F., Barros, M., 2014c. Birnbaum–Saunders statistical modelling: a new approach. Stat. Model. 14, 21–48.

Leiva, V., Saulo, E., Leão, J., Marchant, C., 2014d. A family of autoregressive conditional duration models applied to financial data. Comput. Stat. Data Anal. 79, 175–191.

Leiva, V., Ferreira, M., Gomes, M., Lillo, C., 2015a. Extreme value Birnbaum–Saunders regression models applied to environmental data. Stoch. Environ. Res. Risk Assess. (in press). Available at http://dx.doi.org/10.1007/s00477-015-1069-6.

Leiva, V., Marchant, C., Ruggeri, F., Saulo, H., 2015b. A criterion for environmental assessment using Birnbaum–Saunders attribute control charts. Environmetrics (in press). Available at http://dx.doi.org/10.1002/env.2349

Leiva, V., Santos-Neto, M., Cysneiros, F., Barros, M., 2015c. A methodology for stochastic inventory models based on a zero-adjusted Birnbaum–Saunders distribution. Appl. Stoch. Model. Bus. Ind. (in press). Available at http://dx.doi.org/10.1002/asmb.2124.

Leiva, V., Saunders, S.C., 2015. Cumulative damage models. Wiley StatsRef: Statistics Reference Online. Available at http://onlinelibrary.wiley.com/book/10.1002/9781118445112.

Leiva, V., Tejo, M., Guiraud, P., Schmachtenberg, O., Orio, P., Marmolejo-Ramos, F., 2015e. Modeling neural activity with cumulative damage distributions. Biol. Cybern 109, 421–433.

Leiva, V., Liu, S., Shi, L., Cysneiros, F.J.A., 2016. Diagnostics in elliptical regression models with stochastic restrictions applied to econometrics. J. Appl. Stat. (in press), Available at http://dx.doi.org/10.1080/02664763.2015.1072140.

Lemonte, A., 2013. A new extension of the Birnbaum Saunders distribution. Braz. J. Probab. Stat. 27, 133–149.

Lemonte, A., Cribari-Neto, F., Vasconcellos, K., 2007. Improved statistical inference for the two-parameter Birnbaum–Saunders distribution. Comput. Stat. Data Anal. 51, 4656–4681.

Lemonte, A., Simas, A., Cribari-Neto, F., 2008. Bootstrap-based improved estimators for the two-parameter Birnbaum–Saunders distribution. J. Stat. Comput. Simul. 78, 37–49.

Lepadatu, D., Kobi, A., Hambli, R., Barreau, A., 2005. Lifetime multiple response optimization of metal extrusion die. In: Proceedings of the Annual Reliability and Maintainability Symposium, Institute of Electrical and Electronics Engineers, pp. 37–42.

Lieblein, J., Zelen, M., 1956. Statistical investigation of the fatigue life of deep ball bearing. J. Res. Nat. Bur. Stand. 57, 273–316.

Lin, C.T., Huang, Y.L., Balakrishnan, N., 2008. A new method for goodness-of-fit testing based on type-II right censored samples. IEEE Trans. Reliab. 57, 633–642.

Lio, Y., Park, C., 2008. A bootstrap control chart for Birnbaum–Saunders percentiles. Qual. Reliab. Eng. Int. 24, 585–600.

Lio, Y., Tsai, T., Wu, S., 2010. Acceptance sampling plans from truncated life tests based on the Birnbaum–Saunders distribution for percentiles. Commun. Stat. Simul. Comput. 39, 119–136.

Liu, S., Leiva, V., Ma, T., Welsh, A., 2015. Influence diagnostic analysis in the possibly heteroskedastic linear model with exact restrictions. Stat. Methods Appl. (in press). Available at http://dx.doi.org/10.1007/s10260-015-0329-4.

Macedo, A., 2000. Fatigue of glued laminated wood in toothed amendments. Ph.D. thesis, School of Engineering of São Carlos, University of São Paulo, Brazil.

Malmquist, S., 1950. On a property of order statistics from a rectangular distribution. Scand. Actuar. J. 33, 214–222.

Mann, N., Schafer, R., Singpurwalla, N., 1974. Methods for Statistical Analysis of Reliability and Life Data. Wiley, New York.

Marchant, C., Bertin, K., Leiva, V., Saulo, G., 2013a. Generalized Birnbaum–Saunders kernel density estimators and an analysis of financial data. Comput. Stat. Data Anal. 63, 1–15.

Marchant, C., Leiva, V., Cavieres, M., Sanhueza, A., 2013b. Air contaminant statistical distributions with application to PM10 in Santiago, Chile. Rev. Environ. Contam. Toxicol. 223, 1–31.

Marshall, A., Olkin, I., 2007. Life Distributions. Springer, New York.

Martínez, M., Calil, C., 2003. Statistical design and orthogonal polynomial model to estimate the tensile fatigue strength of wooden finger joints. Int. J. Fatigue 25, 237–243.

Mátyás, L., 1999. Generalized Method of Moments Estimation. Cambridge University Press, New York.

McCarter, K., 1999. Estimation and prediction for the Birnbaum–Saunders distribution using type-II censored samples, with a comparison to the inverse Gaussian distribution. Ph.D. thesis, Kansas State University, Department of Statistics, Kansas.

McCool, J., 1974. Inferential Techniques for Weibull Populations. Aerospace Research Laboratories, Wright-Patterson Air Force Base, Ohio.

McCullagh, P., Nelder, J., 1989. Generalized Linear Models. Chapman and Hall, London, UK.

Meeker, W., Escobar, L., 1998. Statistical Methods for Reliability Data. Wiley, New York.

Michael, J., 1983. The stabilized probability plot. Biometrika 70, 11–17.

Michael, J., Schucany, W., 1979. A new approach to testing goodness of fit for censored samples. Technometrics 21, 435–441.

Michael, J., Schucany, W., Haas, R., 1976. Generating random variates using transformations with multiple roots. Am. Stat. 30, 88–90.

Mills, J., 1997. Robust estimation of the Birnbaum–Saunders distribution. Master thesis, Technical University of Nova Scotia, Nova Scotia, Canada.

Miner, M., 1945. Cumulative damage in fatigue. J. Appl. Mech. 12, A159–A164.

Murthy, V., 1974. The General Point Process. Addison-Wesley, Massachusetts.

Nelson, W., Hahn, G., 1972. Linear estimation of a regression relationships from censored data, part I—simple methods and their applications (with discussion). Technometrics 14, 247–276.

Ng, H., Kundu, D., Balakrishnan, N., 2003. Modified moment estimation for the two-parameter Birnbaum–Saunders distribution. Comput. Stat. Data Anal. 43, 283–298.

Ng, H., Kundu, D., Balakrishnan, N., 2006. Point and interval estimation for the two-parameter Birnbaum–Saunders distribution based on type-II censored samples. Comput. Stat. Data Anal. 50, 3222–3242.

Nocedal, J., Wright, S., 1999. Numerical Optimization. Springer, New York.

Ord, J., 1972. Families of Frequency Distributions. Griffin, London, UK.

Ortega, E., 2001. Influence Analysis in Generalized Log-gamma Regression Models. Ph.D. thesis, University of São Paulo, Brazil.

Ortega, E., Bolfarine, H., Paula, G., 2003. Influence diagnostics in generalized log-gamma regression models. Comput. Stat. Data Anal. 42, 165–186.

Owen, W., 2006. A new three-parameter extension to the Birnbaum–Saunders distribution. IEEE Trans. Reliab. 55, 475–479.

Owen, W., Padgett, W., 1999. Accelerated test models for system strength based on Birnbaum–Saunders distribution. Lifetime Data Anal. 5, 133–147.

Patriota, A., 2012. On scale-mixture Birnbaum–Saunders distributions. J. Stat. Plan. Inference 142, 2221–2226.

Paula, G., 1993. Assessing local influence in restricted regression models. Commun. Stat. Theory Methods 16, 63–79.

Paula, G., 1999. Leverage in inequality constrained regression models. Statistician 48, 529–538.

Paula, G., Leiva, V., Barros, M., Liu, S., 2012. Robust statistical modeling using the Birnbaum–Saunders-t distribution applied to insurance. Appl. Stoch. Model. Bus. Ind. 28, 16–34.

Podlaski, R., 2008. Characterization of diameter distribution data in near-natural forests using the Birnbaum–Saunders distribution. Can. J. For. Res. 18, 518–527.

Press, W., Teulosky, S., Vetterling, W., Flannery, B., 1992. Numerical Recipes in C: The Art of Scientific Computing. Prentice-Hall, London, UK.

R-Team, 2015. R: A Language and Environment for Statistical Computing. R Foundation for Statistical Computing, Vienna, Austria. http://www.R-project.org.

Raftery, A., 1995. Bayesian model selection in social research (with discussion). Sociol. Methodol. 25, 111–196.

Rieck, J., 1989. Statistical Analysis for the Birnbaum–Saunders Fatigue Life Distribution. Ph.D. thesis, Department of Mathematical Sciences, Clemson University, Clemson.

Rieck, J., 1995. Parameter estimation for the Birnbaum–Saunders distribution based on symmetrically censored samples. Commun. Stat. Theory Methods 24, 1721–1736.

Rieck, J., 1999. A moment-generating function with application to the Birnbaum–Saunders distribution. Commun. Stat. Theory Methods 28, 2213–2222.

Rieck, J., 2003. A comparison of two random number generators for the Birnbaum–Saunders distribution. Commun. Stat. Theory Methods 32, 929–934.

Rieck, J., Nedelman, J., 1991. A log-linear model for the Birnbaum–Saunders distribution. Technometrics 3, 51–60.

Rojas, F., Leiva, V., Wanke, P., Marchant, C., 2015. Optimization of contribution margins in food services by modeling independent component demand. Colombian J. Stat. 38, 1–30.

Roussas, G., 1997. A Course in Mathematical Statistics. Academic Press, Massachusetts.

Sanchez, L., Leiva, V., Caro-Lopera, F., Cysneiros, F., 2015. On matrix-variate Birnbaum–Saunders distributions and their estimation and application. Braz. J. Probab. Stat. 29, 790–812.

Sanhueza, A., Leiva, V., Balakrishnan, N., 2008a. A new class of inverse Gaussian type distributions. Metrika 68, 31–49.

Sanhueza, A., Leiva, V., Balakrishnan, N., 2008b. The generalized Birnbaum-Saunders distribution and its theory, methodology, and application. Commun. Stat. Theory Methods 37, 645–670.

Sanhueza, A., Leiva, V., Flores, E., 2009. On a length-biased life distribution based on the sinh-normal model. J. Korean Stat. Soc. 38, 323–330.

Sanhueza, A., Leiva, V., López-Kleine, L., 2011. On the Student-t mixture inverse Gaussian model with an application to protein production. Colomb. J. Stat. 34, 177–195.

Santos-Neto, M., Cysneiros, F., Leiva, V., Ahmed, S., 2012. On new parameterizations of the Birnbaum–Saunders distribution. Pak. J. Stat. 28, 1–26.

Santos-Neto, M., Cysneiros, F., Leiva, V., Barros, M., 2014. On new parameterizations of the Birnbaum–Saunders distribution and its moments, estimation and application. Revstat-Stat. J. 12, 247–272.

Saulo, H., Leão, V., Bourguignon, M., 2012. The Kumaraswamy Birnbaum–Saunders distribution. J. Stat. Theory Practice 1, 1–13.

Saulo, H., Leiva, V., Ziegelmann, F., Marchant, C., 2013. A nonparametric method for estimating asymmetric densities based on skewed Birnbaum–Saunders distributions applied to environmental data. Stoch. Environ. Res. Risk Assess. 27, 1479–1491.

Saulo, H., Leiva, V., Ruggeri, F., 2015. Monitoring environmental risk by a methodology based on control charts. In: Kitsos, C., Oliveira, T., Rigas, A., Gulati, S. (Eds.), Theory and Practice of Risk Assessment. Springer, Switzerland, pp. 177–197.

Saunders, S.C., 1974. A family of random variables closed under reciprocation. J. Am. Stat. Assoc. 69, 533–539.

Saunders, S., 1976. The problems of estimating a fatigue service life with a low probability of failure. Eng. Fract. Mech. 8, 205–215.

Saunders, S., 2007. Reliability, Life Testing and Prediction of Services Lives. Springer, New York.

Seshadri, V., 1999a. The Inverse Gaussian Distribution: A Case Study in Exponential Families. Claredon, New York.

Seshadri, V., 1999b. The Inverse Gaussian Distribution: Statistical Theory and Applications. Springer, New York.

Seto, S., Iwase, K., Oohara, M., 1993. Characteristics of rainfall for a single event. Technical Report No. 93-02. Department of Applied Mathematics, Hiroshima University, Hiroshima, Japan.

Seto, S., Iwase, K., Oohara, M., 1995. Characteristics of rainfall for a single event in Hiroshima City. Tenki (Meteorol. Soc. Jpn) 42, 147–158.

Smith, R., 1991. Weibull regression models for reliability data. Reliab. Eng. Syst. Saf. 34, 55–77.

Soofi, E., Ebrahimi, N., Habibullah, M., 1995. Information distinguishability with application to analysis of failure data. J. Am. Stat. Assoc. 90, 657–668.

St. Laurent, R.T., Cook, R., 1992. Leverage and superleverage in nonlinear regression. J. Am. Stat. Assoc. 87, 985–990.

Therneau, T., Grambsch, P., Fleming, T., 1990. Martingale-based residuals for survival models. Biometrika 77, 147–160.

Tobias, P., 2004. Reliability. In: Group, S.M. (Ed.), e-Handbook of Statistical Methods. NIST/SEMATECH. Available at http://www.itl.nist.gov/div898/handbook.

Tsionas, E., 2001. Bayesian inference in Birnbaum–Saunders regression. Commun. Stat. Theory Methods 30, 179–193.

Tsukatani, T., Shighemitsu, K., 1980. Simplified Pearson distributions applied to air pollutant concentration. Atmos. Environ. 14, 245–253.

Valluri, R.S., 1963. Some recent developments at galci concerning a theory of metal fatigue. Acta Metallurgica 11, 750–775.

Vilca, F., Leiva, V., 2006. A new fatigue life model based on the family of skew-elliptical distributions. Commun. Stat. Theory Methods 35, 229–244.

Vilca, F., Sanhueza, A., Leiva, V., Christakos, G., 2010. An extended Birnbaum–Saunders model and its application in the study of environmental quality in Santiago, Chile. Stoch. Environ. Res. Risk Assess. 24, 771–782.

Vilca, F., Santana, L., Leiva, V., Balakrishnan, N., 2011. Estimation of extreme percentiles in Birnbaum–Saunders distributions. Comput. Stat. Data Anal. 55, 1665–1678.

Villegas, C., Paula, G., Leiva, V., 2011. Birnbaum–Saunders mixed models for censored reliability data analysis. IEEE Trans. Reliab. 60, 748–758.

Volodin, A., 2002. Point estimation, confidence sets, and bootstrapping in some statistical models. Ph.D. thesis, University of Regina, Regina, Canada.

Volodin, I., Dzhungurova, O., 2000. On limit distribution emerging in the generalized Birnbaum–Saunders model. J. Math. Sci. 99, 1348–1366.

von Alven, W.H., 1964. Reliability Engineering. Prentice-Hall, Englewood Cliffs, NJ.

Wang, Z., Desmond, A., Lu, X., 2006. Modified censored moment estimation for the two-parameter Birnbaum–Saunders distribution. Comput. Stat. Data Anal. 50, 1033–1051.

Wanke, P., Leiva, V., 2015. Exploring the potential use of the Birnbaum–Saunders distribution in inventory management. Math. Prob. Eng. Article ID 827246.

Wei, B., Hu, Y., Fung, W., 1998. Generalized leverage and its applications. Scand. J. Stat. 25, 25–37.

Weibull, W., 1951. A statistical distribution function of wide applicability. J. Appl. Mech. 1, 293–297.

Xie, F., Wei, B., 2007. Diagnostics analysis for log-Birnbaum–Saunders regression models. Comput. Stat. Data Anal. 51, 4692–4706.

Zelen, M., Dannemiller, M., 1961. The robustness of life testing procedures derived from the exponential distribution. Technometrics 3, 29–49.

Printed in the United States
By Bookmasters